Arabidopsis thaliana as a Model for Plant-Pathogen Interactions

Edited by
Keith R. Davis and Raymond Hammerschmidt

APS PRESS

The American Phytopathological Society
St. Paul, Minnesota

Cover illustration of life cycle of the *Arabidopsis* pathogen *Peronospora parasitica* courtesy of A. Slusarenko

This publication is based, in part, on presentations from a symposium entitled "*Arabidopsis* as a Model System for Studying Plant–Pathogen Interactions" held in conjuction with the annual meeting of The American Phytopathological Society, August 21, 1991, in St. Louis, Missouri. To make the information available in a timely and economical fashion, this book has been reproduced directly from computer-generated copy submitted in final form to APS Press by the editors of this volume. No editing or proofreading has been done by the Press.

Reference in this publication to a trademark, proprietary product, or company name by personnel of the U.S. Department of Agriculture or anyone else is intended for explicit description only and does not imply approval or recommendation to the exclusion of others that may be suitable.

Library of Congress Catalog Card Number: 93-71322
International Standard Book Number: 0-89054-153-1

Printed in the United States of America on acid-free paper

The American Phytopathological Society
3340 Pilot Knob Road
St. Paul, Minnesota 55121-2097, USA

CONTENTS

PREFACE

Over the past several decades, a vast literature has accumulated describing the interaction of plants with pathogenic microorganisms. It has become apparent from these studies that plants are quite efficient at recognizing potential pathogens and mounting an effective defense. Genetic analyses of a number of plant-pathogen combinations indicate that in some cases, a relatively simple gene-for-gene interaction between specific resistance genes in the plant and a corresponding avirulence gene in the pathogen determines whether the interaction will result in a resistance response or disease. Although substantial progress has been made in the identification and characterization of pathogen avirulence genes, not a single plant resistance gene corresponding to any of these avirulence genes has been isolated. Other studies using biochemical and molecular approaches have shown that a number of inducible responses are correlated with disease resistance. However, since the overall response of the plant to microbial attack is complex, it has been difficult to clearly demonstrate by these correlative approaches that a particular response is required for resistance. In addition, although significant progress has been made in identifying specific genes that are activated during attempted infection, these studies have not provided much detailed information concerning the mechanisms involved in the activation and coordinated regulation of these putative defense genes.

Studies using biochemical and molecular approaches combined with genetic analyses of plants with defense response mutations would provide a very powerful method for dissecting the complex pattern of gene expression that is associated with plant disease resistance. However, few of plant models that have been used to date for molecular studies of disease resistance are amenable to genetic approaches. Over the past few years, several groups have been investigating the feasibility of using *Arabidopsis thaliana* as a model host plant for studies of disease resistance. These groups have made rapid progress in identifying potential pathogen systems and characterizing the genetic parameters influencing the interaction *A. thaliana* with bacterial, fungal, and viral phytopathogens. These initial studies provide a strong indication that *A. thaliana* is indeed a useful system for studying plant disease resistance which will undoubtedly contribute new information concerning the recognition of potential pathogens and the activation of a resistance response.

In an attempt to introduce the development of *A. thaliana* as a model host to the plant pathology community as a whole, a symposium was organized for the 1991 Annual Meeting of APS that brought together some of the researchers who are actively pursing studies of the interaction of *A. thaliana* with a variety of plant pathogens. This symposium was organized by K. R. Davis of the APS Committee on Biochemistry, Physiology, and Molecular Biology and was supported by funds provided by APS and E. I. Du Pont De Nemours & Company. This symposium volume was edited by K. R. Davis and R. Hammerschmidt with the assistance of APS Press senior editor J. Fletcher. We thank the authors for their contributions to this volume and are grateful for the excellent editorial assistance provided by P. Gerds and M. Marais. We also thank Dr. P. A. Scolnik for his efforts in obtaining financial support for this symposium and the anonymous reviewers who provided many useful comments concerning this volume.

Keith R. Davis
The Biotechnology Center,
Department of Plant Biology,
and Department of Plant Pathology
The Ohio State University
Columbus, OH 43210

Raymond Hammerschmidt
Department of Botany and Plant Pathology
Michigan State University
East Lansing, MI 48824

ARABIDOPSIS AS A MODEL PLANT SYSTEM

Keith R. Davis

Departments of Plant Biology and Plant Pathology and the Biotechnology Center, The Ohio State University, Columbus, OH 43210-1002

Arabidopsis thaliana is a small cruciferous plant that has become an increasingly important model system for investigating many aspects of higher plant biology. *A. thaliana* was first introduced as an experimental system by Laibach in Germany (92) in the 1940's and was later refined by Reinholz and Langridge (95,125). Since that time, *A. thaliana* has proven to be an extremely useful system for both qualitative and quantitative genetics (reviewed in 63,89,112,113,124,141).

This small weed offers many experimental advantages, including small size, short life cycle (4-6 weeks from seed to seed), self-fertility, and the ability to produce large amounts of seed (10,000 or more per plant). These characteristics allow for the rapid growth and analysis of a large number of individuals in a minimum of space and subsequent rapid amplification of useful genotypes for further study. In addition, *A. thaliana* has a compact genome (70,000 to 100,000 kb per haploid genome) that contains only about 10% highly repetitive sequences (96). This makes *A. thaliana* an ideal organism for genetic and molecular genetic studies such as mutant analysis, the molecular cloning of genes, and the construction of a detailed physical map of its genome. Another important feature of *A. thaliana* is that it can be easily genetically engineered using either *Agrobacterium*-mediated transformation (106,158), direct DNA uptake by protoplasts (16) or particle bombardment of tissues (136). The ability to genetically transform *A. thaliana* allows for detailed analyses of gene function and expression. In addition, it allows for the functional complementation of a mutant phenotype with cloned genes, the critical final step in isolating genes by mutational analysis.

MUTANT ANALYSIS IN *ARABIDOPSIS*

Of the many advantages *A. thaliana* offers as an experimental system, one of the most attractive is the ability to clone genes that have been identified by mutational analysis. The combined efforts of several laboratories have resulted in the generation of detailed genetic maps of the *A. thaliana* genome. These genetic maps include morphological and biochemical markers (88), as well as molecular markers such as restriction fragment length polymorphisms (RFLPs) and random amplified polymorphic DNAs (RAPDs). Two genetic maps based on RFLP markers have been developed, comprising about 380 mapped probes (9,119). A map based on 60 of these RFLP markers and an additional 252 RAPD markers has also been recently described (126). Recent calculations indicate that approximately 90% of the total *A. thaliana* genome lies within 0.8 Mbp of an RFLP marker and that almost 50% is within 0.27 Mbp of a RFLP marker (38,119).

In addition to the development of genetic maps, a major effort is underway to develop a complete physical map of the *A. thaliana* genome consisting of overlapping sets of cosmids and yeast artificial chromosomes

(YACs). The cosmid map being developed by Goodman and coworkers currently covers 90-95% of the *A. thaliana* genome and includes over 17,000 cosmids organized into 750 sets of overlapping clones (67). Physical maps based on sets of overlapping YACs are also being developed (38,64,73,162). As these complementary physical maps are completed, researchers will have available an extremely detailed description of the *A. thaliana* genome and a set of powerful tools with which to clone genes that have been identified only by mutations.

Two other powerful tools have been developed for mutant analysis in *A. thaliana* that complement map-based cloning methods. The first approach, T-DNA tagging, relies on using "seed transformation" or transformation of zygotic embryos with *Agrobacterium* as methods that allow T-DNA to be used as a mutagen. T-DNA-tagging has already been used to isolate several genes (41,44,45,105,168) and will undoubtedly be increasingly useful as larger populations of transformed plants with low levels of somaclonal variation become available. A second recently developed approach is a genomic substraction method that can be used to clone genes corresponding to deletion mutations (144). This method has been successfully used to clone the *A. thaliana ga-1* locus (145) and should become more useful as methods for efficiently isolating deletion mutants are established.

CONCLUSIONS

Genetic approaches have proven to be extremely powerful in the dissection of complex developmental, biosynthetic and regulatory pathways in a number of model systems such as *E. coli, D. melanogaster,* and *C. elegans.* Recent studies of *A. thaliana* development and metabolism (42,108) clearly demonstrate that this small weed is amenable to the types of molecular and biochemical genetic methods that would be ideal for investigating the complex phenotype of disease resistance (22,23). As the *A. thaliana*-pathogen systems described in subsequent chapters become more developed, we can expect that the application of genetic approaches will provide new and important insights into our understanding of the recognition of microorganisms by plants and the establishment of an effective defense.

THE GENETIC BASIS OF RESISTANCE OF *ARABIDOPSIS THALIANA* L. HEYHN. TO *PERONOSPORA PARASITICA*

Brigitte Mauch-Mani, Kevan P.C. Croft & Alan Slusarenko*, Institut für Pflanzenbiologie, Zollikerstr. 107, CH-8008 Zürich, Switzerland

*To whom correspondence should be addressed

SUMMARY

The inheritance of resistance to *Peronospora parasitica* in the Columbia and RLD (also known as Rschew) accessions of *Arabidopsis thaliana* was studied by setting up crosses between resistant and susceptible parents and following segregation of the resistance phenotype in F_2 populations. Resistance appears to be conditioned by single dominant genes in both accessions and this was confirmed in backcrosses of F_1 individuals with a susceptible parent. The resistance phenotypes associated with Columbia and RLD, respectively, were slightly different in that although both showed hypersensitive resistance, the pathogen made more ingress into Columbia than RLD. At present it is not possible to say whether the resistance gene in Columbia is allelic to that in RLD or whether two linked or unlinked loci are involved. There was no evidence for maternal inheritance of resistance in either ecotype. There was evidence of a gene dosage effect for resistance in Columbia as it was possible to recognize the heterozygotes which, although still resistant, showed an intermediate phenotype.

Biochemical changes in infected tissues were also investigated and data are presented showing an increase in chitinase enzyme activity and transcripts of the basic, but not the acidic form of *Arabidopsis* chitinase associated with the incompatible interaction. Peroxidase enzyme activity showed a similar pattern of increase, but no significant pattern of infection-related changes was observed for superoxide dismutase, catalase, ascorbate peroxidase, lipolytic acyl hydrolase, lipoxygenase and a linoleic acid-13-hydroperoxide decomposing activity. The potential role of ethylene as a signal in plant defense responses in *Arabidopsis* was also investigated.

Systemic acquired resistance induced by chemical treatment with the plant immunization substance dichloroisonicotinic acid (INA) was also observed. Treatment with as little as 0.1 ppm INA prior to inoculation reduced markedly tissue colonization and sporulation by the pathogen.

INTRODUCTION

Arabidopsis thaliana is a small crucifer which has long been recognized as a suitable subject for classical plant genetic studies (92). The plants have a short generation time (9-12 weeks), are normally self-fertilizing but easy to cross, and produce prolific amounts of seed. The small genome size (currently estimated to be approx. 100,000 kbp) and low amount of repetitive DNA have made *Arabidopsis* an attractive subject for investigations into the molecular genetics of plant development and several aspects of metabolism and cell biology (see 113 for a recent review). In *Arabidopsis* it is theoretically quite easy to isolate a genetic locus which has a recognizable phenotype by a map-based cloning strategy (141) by the co-segregation of known genetic or RFLP markers with the locus of interest, coupled with chromosome walking (169). Since *Arabidopsis* can be transformed by *Agrobacterium* and transgenic plants regenerated, the identity of cloned loci can be tested by functional complementation. Thus, for the first time it is possible to clone and identify a genetic locus with a recognizable phenotype without the need for it to be tagged with a transposon, and without having a detailed knowledge of the biochemical function of the gene product.

The use of *Arabidopsis* to gain an insight into the molecular biology of plant-pathogen interactions has been hindered by the lack of well characterized pathosystems (86). However, in recent years several fungal, bacterial and viral pathogens of *Arabidopsis* have been described (see this volume). For the record we would also like to mention two further fungi which we have isolated from naturally infected plants, firstly, the oomycete *Pythium* (Fig. 1a) and secondly the deuteromycete *Chromelosporium* (Fig. 1b) which has *Peziza* as the teleomorphic (perfect) stage (69). *Pythium* infection was observed as damping-off of *Arabidopsis* seedlings in the glasshouse. The fungus was isolated in axenic culture and used to reinfect healthy *Arabisopsis* plants by applying a mycelial plug to the leaves and keeping the plants under high humidity. Within three days the plants were extensively colonized by the fungus and showing soft rot symptoms. Numerous oospores were present in infected leaves. The pathogen was tentatively identified as *Pythium paroecandrum* by Dr. G. White (HRI, Wellesbourne, UK). *Chromelosporium* infection occurred from inoculum in the potting compost.

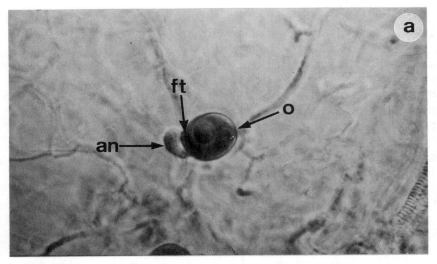

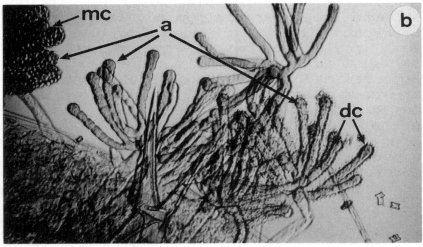

Figure 1. a: Sexual reproduction of *Pythium paroecandrum* in a leaf of *Arabidopsis*. A globose oogonium (o) can be seen with a club-shaped antheridium (an) in the process of forming a fertilization tube (ft).
b: Conidiophores of *Chromelosporium* with ampullae (a) at different developmental stages; some bearing mature conidia (mc) and some developing conidia (dc).

The orange colored fungus grew from the soil surface into the plants and sporulated from the petioles and lower side of the leaves. Infected leaves eventually became necrotic.

However, the recently described downy mildew infection of *Arabidopsis* (87) so far shows the greatest promise as a model pathosystem not only to clone classical resistance genes, but also to investigate the molecular biology of resistance mechanisms (140). In order to define targets for manipulation at the molecular level, it is first necessary to characterize the resistance-associated changes in biochemistry of challenged leaves and to this end, we have investigated several enzymes whose activities are thought to be implicated in resistance responses of other plants.

GENETICS OF RESISTANCE

Landsberg erecta (susceptible) x Columbia (resistant)

The isolate of *P. parasitica* used in this work is called the WELA isolate (found in WEiningen near Zürich and virulent on LAndsberg erecta). Sixty F_1 plants were tested and all were found to be resistant to the WELA isolate of *Peronospora parasitica*, showing that resistance was dominant to susceptibility. Since the susceptible accession Landsberg erecta was used as the female parent, the possibility that self-fertilization rather than out-crossing had occurred was negated and resistance was clearly not maternally inherited. The F_1 plants were allowed to self and resistance in the F_2 population segregated in a 3:1 ratio resistant:susceptible. Thus, one can conclude that the resistance phenotype in Columbia is conditioned by a single dominant locus (Fig.

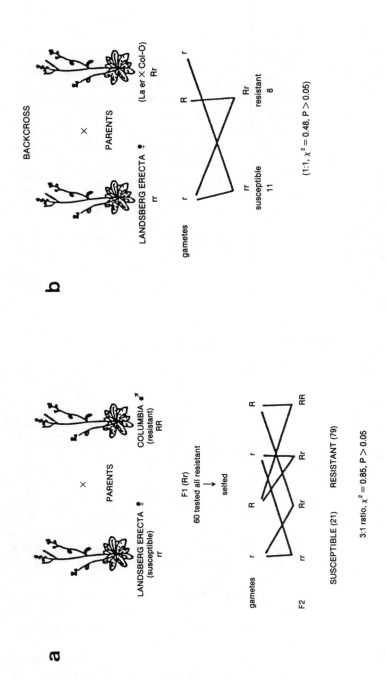

Figure 2. Inheritance of resistance to *Peronospora parasitica* in a: the cross between Landsberg erecta (La er) and Columbia (Col-O) and b: the backcross between the F_1 and Landsberg erecta.

2a). This conclusion is supported by the (1:1) segregation ratio in backcrosses of F_1 individuals with susceptible parents (Fig. 2b). The backcrosses were set up as a diallel cross with both combinations of male and female parents and in both cases a 1:1 ratio was seen for resistant:susceptible; this serves as a further check that maternal inheritance is unimportant and that gamete-specific genomic imprinting effects do not seem to be occurring.

In scoring the F_2 population microscopically, two classes of resistant plants were observed. One class showed a hypersensitive reaction, virtual absence of mycelium, no oospores and no conidiophores (*i.e.* the parental resistance phenotype). The second phenotype although generally similar had local clusters of oospores in small patches in individual leaves. Twenty-eight individuals of the F_2 had the parental phenotype, 51 the second phenotype and 21 were completely susceptible and supported profuse asexual sporulation and accumulated many oospores in the infected leaves. We interpreted these numbers as suggesting that the "intermediate" class was heterozygous for the resistance allele (*i.e.* Rr). The chi^2 test for a 1:2:1 ratio RR:Rr:rr gives a value of 1.02, 2 degrees of freedom, P>0.05. However, assigning an individual to the heterozygote class is necessarily tentative because of the range of resistance phenotypes seen in Columbia (see later).

Weiningen (susceptible) x RLD (resistant)

All of 100 F_1 plants tested from a cross between ecotypes Weiningen (susceptible) and RLD (resistant) were resistant to infection with the WELA isolate of *Peronospora parasitica*. This shows that resistance in RLD is a dominant trait and the 3:1 segregation observed in the F_2 population 5-7 days after inoculation indicates that it is probably conditioned by a single dominant gene (representative data for an F_2 population was 38 resistant and 12 susceptible plants; chi^2 for a 3:1 ratio = 0.014, P>0.05). Resistance segregated 1:1 with susceptibility in the progeny from backcrosses of the F_1 with susceptible Weiningen, thus lending support to this contention (representative backcross: 6 resistant, 8 susceptible, chi^2 for a 1:1 ratio = 0.28, P>0.05). Interestingly, using a detached leaf pathogenicity assay and scoring 24 h after inoculation, the proportion of resistant individuals in the F_2 was higher than expected if resistance were conditioned by a single, dominant locus (140). This observation might indicate that genes of "minor" effect have a retarding influence on the pathogen early on in infection of RLD but that they are overcome in the absence of the "major" resistance gene.

Different isolates of the pathogen have been found which show pathotypic variation on several different A. *thaliana* ecotypes (17, Holub *et al*. this volume). Of particular interest in this case is the isolate EMOY2 which is virulent on Columbia but avirulent on Landsberg erecta. The EMOY2 and WELA (and CALA) isolates show the inverse pattern of host specificity with respect to the ecotypes Columbia and Landsberg erecta. Thus, the reciprocal check criterion, suggesting a gene for gene type of interaction between host and pathogen (159), is satisfied for the interaction between A. *thaliana* and P. *parasitica*.

The finding that WELA is virulent on Landsberg erecta but avirulent on Columbia, and that resistance in Columbia is conditioned by a single dominant locus, is a very good pre-requisite for a map-based cloning strategy to isolate the resistance locus (169). These two accessions, along with Niedersenz, were used in crosses to define the published linkage maps of *A. thaliana* (9, 119). Thus, individual RFLP and genetic probes known to highlight polymorphisms between these two parents can be used directly to follow cosegregation with the resistance locus. In the case of Weiningen and RLD each probe needs to be tested to see if a polymorphism is highlighted between these parents. Since it appears we can recognize the heterozygote between Columbia and Landsberg erecta, this will also facilitate mapping in segregating populations. Mapping is at present in progress using DNA extracted from F_3 families from F_2 individuals.

The observation that, although resistance in the Columbia x Landsberg erecta cross is dominant, this dominance is incomplete, i.e. there is evidence of a gene dosage effect and the heterozygote is recognizable, has interesting implications for resistance gene function. Incomplete dominance has also been reported in the *Bremia lactucae/Lactuca sativa* interaction (Crute 1987), where increasing incompatibility followed the trend RR/AA>RR/Aa>Rr/AA>Rr/Aa for the different genotype combinations. If, as is commonly held, the direct or indirect products of resistance genes are receptors for signal molecules from the pathogen, which when triggered act as a switch to activate the response genes needed for active defense, then incomplete-dominance would not necessarily be expected. Put simply, how can you half switch something on? The implications of this observation on speculations about resistance gene function have been discussed by Crute & Norwood (15).

Phenotypic Differences of Resistance in Columbia and RLD

There is a higher level of resistance in accession RLD compared to Columbia. In general, the infection court is larger in Columbia with a few cells responding hypersensitively and hyphae growing to a few cell diameters from the penetration point. However, in some infections larger necrotic areas containing a few oospores were observed and, very rarely, sparse conidiophores developed on infected Columbia individuals. A single infected individual could show the whole range of symptoms. Where asexual sporulation was observed, this seemed to occur in zones on younger leaves in plants which on other leaves showed the more resistant phenotype. The reaction did not resemble a classical susceptible phenotype, however, as a band of necrotic host cells in the wake of hyphal growth could be traced from the penetration site to the point of emergence of one to three conidiophores. Thus, the resistance locus in Columbia to the WELA isolate of *Peronospora parasitica* could be described as somewhat "leaky." In contrast, the HR in RLD is usually confined to one or two cells at the site of penetration and neither sexual nor asexual sporulation of the pathogen has been observed on this host. The different resistance phenotypes of RLD and Columbia are reminiscent of the effects of different resistance gene/avirulence gene combinations in the interaction of *Bremia lactucae* with lettuce (*Lactuca sativa*; 167). However, another possibility is that the same resistance gene is present in both cultivars but that genetic background plays a role in conditioning the phenotype. In order to gain more information on this point we tested an F_3 population derived from about 60 F_2 plants from a cross between RLD and Columbia to see if individuals susceptible to WELA could be found. If the same gene were involved in RLD and Columbia we would expect no segregation for resistance in the F_3; any segregation would show that two genes, or more correctly two different loci, were involved and would indicate the degree of linkage. We found 40 clearly susceptible individuals out of 1,600 F_3 plants tested, thus indicating that resistance to WELA in accessions RLD and Columbia is conditioned at different loci. In addition, we are introgressing the RLD and Columbia resistance traits into Weiningen to produce near-isogenic lines. If these maintain the RLD and Columbia resistance phenotypes, then one could conclude that genetic background effects were not important.

Interestingly, we see differences in the degree of susceptibility of Weiningen and Landsberg erecta to the WELA isolate of *P. parasitica*. Colonization is more rapid with Weiningen and asexual sporulation occurs by 7 days after inoculation, whereas with Landsberg erecta colonization is slower and asexual sporulation is less heavy until about 10 days after infection. Holub and Crute (in 17) also report variation in the degree of susceptibility of *A. thaliana* accessions to *Peronospora* isolates and state that ecotype Weiningen, while not always their most susceptible host, is so far a universal suscept. It would be most interesting to attempt to rank susceptible accessions in their degree of resistance and pathogen isolates according to their virulence and gain an insight into the genetic basis of this variation. Thus, it may be possible to use *Arabidopsis* as a model to clone loci involved in horizontal resistance to *P. parasitica*, in addition to cloning loci involved in vertical resistance in a gene-for-gene type interaction (137, 159).

BIOCHEMICAL RESPONSES IN INFECTED TISSUES

Arabidopsis is not an easy subject for biochemical studies. We found that enzyme extraction procedures which gave good preparations with other plants were ineffective. Routinely, tissue was first ground in liquid nitrogen and then homogenized (1 g in 4 ml) on ice in 50 mM Tris-HCl (pH 8) containing 0.2% Triton X-100, 10% (v/v) glycerol, 10 mM Na_2EDTA, 100 mM KCl, 20 mM sodium metabisulphite and 1 mM benzamidine-HCl. The slurry was filtered through two layers of muslin and the filtrate centrifuged at 10,000 x *g* for 10 min at 4°C to remove the remaining debris. The supernatant was then desalted on PD10 columns (Pharmacia), eluted with 50 mM potassium phosphate buffer (pH 7), and the eluate made 10% (v/v) with glycerol, aliquoted and frozen. The above extraction procedure did enable us to measure at least basal levels of some enzymes but changes upon infection were not always apparent. Thus, no significant pattern of change on infection was observed for superoxide dismutase, catalase, ascorbate peroxidase, lipolytic acyl hydrolase, lipoxygenase and a linoleic acid-13-hydroperoxide decomposing activity. Perhaps in *Arabidopsis* these enzymes have no role in resistance or, because only relatively few cells show a hypersensitive reaction, any changes are diluted out in comparison with the vast bulk of non-stimulated cells in the leaf. A further complicating factor for several enzymes was the different basal levels of specific activity between accessions and plants of different age.

Chitinase

In a comparison between RLD (resistant) and Weiningen (susceptible), chitinase enzyme activity measured using a tritiated chitin

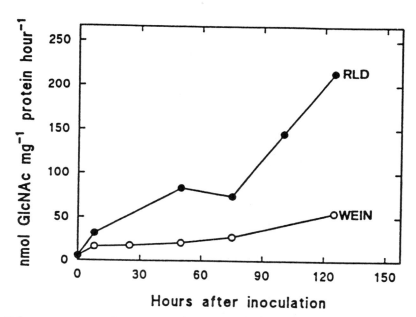

Figure 3. Changes in chitinase activity after inoculation with *P. parasitica* (WELA).

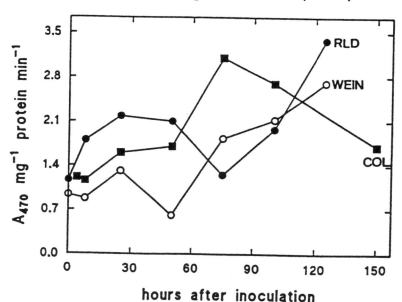

Figure 4. Changes in peroxidase activity after infection with *P. parasitica* (WELA).

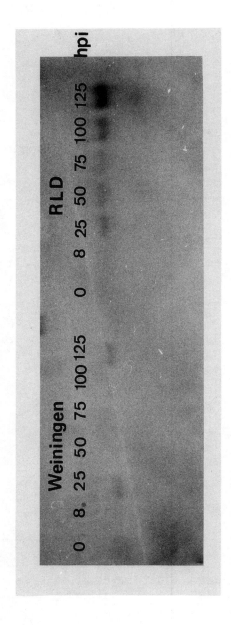

Figure 5. Differential accumulation of basic chitinase transcripts in *Arabidopsis* accessions RLD (resistant) and Weiningen (susceptible) after inoculation with the WELA isolate of *Peronospora parasitica*. hpi = hours after inoculation.

18

substrate increased earlier and to a greater degree in the incompatible than in the compatible combination (Fig. 3). This increase correlated with accumulation of basic chitinase transcripts (Fig. 5) but no induction of acidic chitinase could be detected (classification according to 132). However, it appears that the acidic chitinase promoter is activated by *Rhizoctonia* infection (133). Interestingly, we found that ethylene at 100 ppm did not appear to induce chitinase enzyme activity in Weiningen although ethylene has been reported to induce transcripts of the basic chitinase in Columbia (132).

Since *P. parasitica* is an oomycete and would therefore be expected to have a predominantly cellulose-based cell wall, chitinase may not play an important role in resistance to this pathogen.

Peroxidase

Peroxidase enzyme activity, measured using guaiacol as a substrate, increased earlier in RLD and Columbia (resistant) than in Weiningen (susceptible; Fig. 4). In Columbia peroxidase activity increased later than in RLD. This might reflect the apparently less intense resistance phenotype observed at the microscopic level in this accession (see earlier). Interestingly, ethylene treatment caused a doubling of peroxidase activity by 16 h after treatment.

INDUCED RESISTANCE

Weiningen and Landsberg plants, normally susceptible to infection with the WELA isolate of *P. parasitica*, showed reduced colonization and sporulation by the pathogen after treatment with as little as 0.1 - 1.0 ppm 2,6-dichloroisonicotinic acid (INA, kindly supplied by CIBA-GEIGY, Basel, CH) as a soil drench. INA has not been shown to have antifungal activity in its own right (111) and is presumed to exert its effect by priming endogenous resistance mechanisms in the plant. Indeed, INA treatment is associated with the induction of genes coding for different classes of PR proteins, the accumulation of which correlates well with the timing of resistance induction (157).

ACKNOWLEDGMENTS

Thanks are due to the Kanton of Zürich and the Royal Society of London for financial support, to Debbie Samac for the gift of *Arabidopsis* acidic and basic chitinase clones, to Ian Crute and Eric Holub for a generous gift of F_2 seed from a Columbia x RLD cross and also for critical reading of the manuscript, and to Dr. G. White for identifying the *Pythium* isolate at the species level.

IDENTIFICATION AND MAPPING OF LOCI IN *ARABIDOPSIS* FOR RESISTANCE TO DOWNY MILDEW AND WHITE BLISTER

Eric Holub[1], Ian Crute[1], Edemar Brose[2] and Jim Beynon[2]

[1]Horticulture Research International-East Malling, West Malling, Kent, UK.

[2]Plant Biochemistry Department, Wye College, University of London, Wye, Kent, UK.

Although the study of interactions between *Arabidopsis thaliana* and fungal pathogens is still in its infancy, this wild crucifer will clearly play an important role in advancing our understanding of how a plant recognizes a pathogen and defends itself. Opportunities will also emerge from the study of wild pathosystems which reach beyond understanding interactions at the molecular and whole organism level to elucidate interactions between populations and reveal processes of host/pathogen co-evolution. Prospects for the integration of research at these various levels of complexity will depend, among other things, upon the characterization of natural phenotypic and genetic variation which exists in a given pathosystem, and upon the availability of molecular markers linked closely to resistance loci. With this in mind, two biotrophic fungal pathogens of *Arabidopsis* are described which are well suited for programs of field and laboratory based research.

21

DOWNY MILDEW AND WHITE BLISTER IN THE UK

Natural infection of *A. thaliana* by the biotrophic oomycete fungi *Peronospora parasitica* and *Albugo candida* was observed by Drs. Paul Williams (Plant Pathology Department, University of Wisconsin, Madison) and Bill Barlow (former Head, Plant Physiology Department, East Malling Research Station) in Kent, UK during an expedition in spring 1987. Their finding provided the basis for our research. Since then, we have collected numerous fungal isolates and have developed experimental methods thereby allowing us to study natural genetic variation that exists in these two pathosystems. The objective at present is to map and ultimately to clone genes which control resistance to the fungi.

Both fungi are common pathogens of *A. thaliana* in the UK and often grow intimately together in the same host tissue (Figure 1). In two consecutive springs (1990-91), thirty-two *A. thaliana* populations in Kent were examined. Of these, 19% contained plants infected by *P. parasitica* alone, 19% by *A. candida* alone, and 9% contained plants infected with both fungi. Neither fungus was found in the remaining 53% of populations. Infection was most frequently observed as profuse asexual sporulation at the undersurface of rosette leaves. Infection of stems and petioles was also observed. Most infected plants were bolting and were clearly capable of going on to produce seed. If infection by these fungi is debilitating to *A. thaliana*, it is likely to result in poor competitive ability at the seedling stage or reduced fecundity of flowering plants.

P. parasitica and *A. candida* from *A. thaliana* are morphologically similar to isolates of the same species collected from other crucifers such as brassicas or

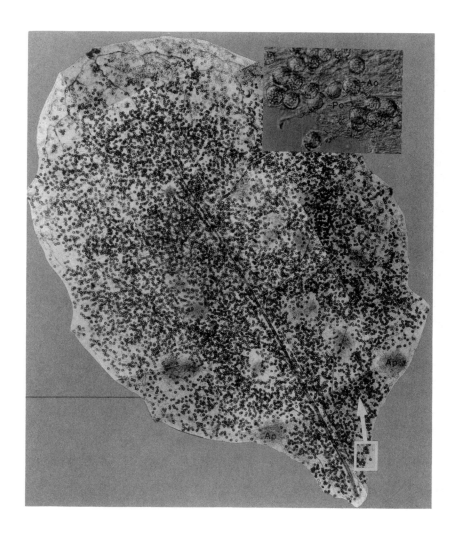

Figure 1. Sexual reproduction of *Albugo candida*
(isolate EMAL) and *Peronospora parasitica* (isolate
CALA1) in a glabrous rosette leaf of *Arabidopsis
thaliana* (accession Col0 *gl-1*). An area near the
petiole shows tuberculate oospores of *A. candida*
(Ao) and smooth-walled oospores of *P. parasitica*
(Po). Pustules of *A. candida* (Ap) are also visible
here as granular patches.

Capsellabursa pastoris. However, from preliminary observations it appears that isolates from *A. thaliana* are only pathogenic on this or other species of *Arabidopsis*.

COTYLEDON ASSAY

A cotyledon assay has been used to examine phenotypic variation among accessions for response to different pathogen isolates of both fungi (17). Seedlings are inoculated seven days after sowing by placing a 1 µl drop containing c. 50 spores on each cotyledon. With *P. parasitica*, conidiosporangia are used for inoculum and plants are assessed 3 and 7 days after inoculation (dai) for the production of sporophores and macroscopic host reaction. With *A. candida*, zoospores are used and the interaction is assessed 10 dai.

PHENOTYPIC VARIATION FOR RESPONSES TO
P. parasitica

Initially, interaction phenotypes were classified simply by recording the presence or absence of asexual sporulation 7-10 dai. Two mass sporangial isolates of *P. parasitica* (EMOY1 and CAND1) were used to test numerous accessions of *A. thaliana* that had been collected from locations worldwide. The accessions were classified into four phenotypic groups (Table 1) including examples where both isolates (S/S), neither isolate (N/N), or only one or the other isolate sporulated (N/S and S/N). Four additional isolates were used subsequently to inoculate the same accessions and all six isolates proved to be six different pathotypes.

Phenotypic variation can exist within as well as between populations of both the plant and the fungus. Three patterns of reaction to isolates EMOY1 and CAND1 (S/S, S/N and N/S) were found among progeny of *A. thaliana* plants collected from the grounds of the research station at East Malling (Table 1). Isolates of different patho-types have also been collected from the same location. For example, the accession Col0 reacted differently to two isolates from East Malling (EMOY1 and EMOY2, Table 1). Both isolates were derived from dried leaf tissue containing oospores (17). The leaves were collected from plants growing in a 1 m^2 area. Similarly, La-er responded differently to two mass-sporangial isolates (CAND1 and CALA1) collected from the same garden in Canterbury.

More detailed characterization of phenotypic variation has been possible by comparing the amount of sporulation and the type of host lesion (Table 2). For example, A. *thaliana* accessions Oy0 and Col0 react differently to P. *parasitica* isolate EMOY2. Oy0 is especially susceptible to this isolate; sporophores begin to emerge three days after inoculation with no obvious host response. By contrast, the reaction of Col0 typically involves collapse of mesophyll cells and hyphae, and sporophores emerge 5-7 dai. This phenotypic difference may be under the control of an allele at a single locus which conditions incomplete resistance in Col0 to EMOY2 (see F$_2$ Col0 x Oy0, Table 3).

Three contrasting types of host lesion have been found among reactions of accessions Nd0, La-er and Ws0 to isolate EMOY2. On La-er, this isolate causes minute necrotic flecks which only become clearly visible to the naked eye 7 dai. However, much larger necrotic pits develop on Nd0 and Ws0. These pits are discrete epidermal lesions up to 1 mm in diameter by 3 dai and are often surrounded by a chlorotic halo. On Nd0, the pits usually continue to expand until the cotyledon is entirely necrotic. The lesion never progresses further so that a seedling responding in this way will continue to grow.

Table 1. Phenotypic responses of *Arabidopsis thaliana* to *Peronospora parasitica* and *Albugo candida*.

Geographic Origin of A. thaliana	P. parasitica[a]				A. candida[a]	
	CAND1	CALA1	EMOY1	EMOY2	EMAL	KESK
Cape Verde Island (Cvi0)	S[b]	S	S	S	S	nt
Japan (Tsu0)	S	S	S	S	S	S
Switzerland (Wein)	S	S	S	S	S	nt
Germany (Nd0)	S	S	N	N	S	S
Spain (Pla0)	S	S	S	N	S	S
Norway (Oy0)	N	N	S	S	S	S
Poland (Mh0)	N	N	S	N	S	S
USA (Tul0)	S	S	S	S	S	nt
Austria (Pi0)	S	S	N	N	S	S
Finland (Te0)	N	N	N	N	S	nt
Libya (Mt0)	N	N	N	N	S	S
Netherlands (Hi0)	N	N	N	N	S	S
Germany (La-er)	N	S	N	S	S	S
Germany (Col0)	N	N	N	N	S	S
Russia (RLD)	N	N	N	N	S	S
UK (E. Malling)	S	S	S	S	S	S
	N	N	N	N	S	S
	S	S	S	S	S	S
UK (Keswick)	N	N	N	N	N	S
	S	S	S	S	N	S
	S	nt	S	nt	S	S

a *P. parasitica* isolates were all collected in Kent, UK from locations near Canterbury (CAND1,CALA1) and East Malling (EMOY1,EMOY2). *A. candida* isolates were collected from UK locations near East Malling (EMAL) and Keswick (KESK).

b Accessions were classified according to presence (S) or absence (N) of profuse asexual sporulation 7-10 days after inoculation. nt= non-tested.

Table 2. Phenotypic variation among *Arabidopsis thaliana* accessions for response to isolates of *Peronospora parasitica*.

| | *A. thaliana* accession | | | | | | |
Isolate	Col0	Nd0	Oy0	Laer	Wein	RLD	Ws0
EMOY2	DFL[a]	NPX	EH	NF	DL	NF	NPT
CALA1	RF	EH	NF	EH	DL	NF	NCT

[a] Description of fungal growth includes timing of sporophore emergence as early (E) 3 days after inoculation (dai), delayed (D) at least 48 h, or none observed (N); and also the intensity of sporulation as heavy (H), low to moderate (L), or a rare sporophore (R). Plant reaction was characterized as minute flecking lesions (F) 7 dai, chlorotic spots (C) 3 dai becoming necrotic 7 dai, and necrotic pitting (P) 3 dai. Lesions were also characterized 7 dai as determinate (T) or expanded (X) relative to size of lesions 3 dai.

In contrast, on Ws0 the pits are determinate because necrosis rarely expands after 3 dai. Furthermore, Ws0 responds to infection by CALA1 with a similar determinate lesion, except deep pits seldom form. Hence, we expect to find two alleles in Ws0, one which causes pronounced pitting reaction to EMOY2 and a second which causes a determinate lesion in reaction to either isolate. Determinate pits result when both alleles are expressed.

Microscopic differences have also been observed between the flecking necrosis of Oy0 and Col0 following inoculation with isolate CALA1. Necrotic lesions on Oy0 involve less than ten host cells and the fungal hyphae are seldom visible beyond the lesion. Whereas, necrotic lesions on Col0 involve more than twenty cells, hyphal extension is often

Table 3. Segregation for response to *Peronospora parasitica* among F2 progeny from crosses between *Arabidopsis thaliana* accession Col0 and five other accessions (selfed progeny response for each accession is also shown).

Isolate	Female Parent	Male Parent					
		Col0	Nd0	Oy0	La-er	RLD	Wein
EMOY2	Col0	0:10:40:1	65:12:14:6	0:0:62:39	0:51:10:0	0:34:2:1	0:2:52:3 [a]
	Nd0		112:1:0:0				
	Oy0			0:1:11:113			
	La-er				0:39:0:0		
	RLD					0:22:0:0	
	Wein						0:13:12:9
CALA1	Col0	0:253:7:0	0:21:50:9	0:318:9:0	0:16:54:23	0:71:1:0	0:9:37:12
	Nd0		0:0:16:197				
	Oy0			0:258:0:0			
	La-er				0:0:8:70		
	Wein						0:7:10:9
	RLD					0:24:0:0	

[a] Numbers of plants in each of four phenotypic classes (left to right): pitting necrosis of cotyledons 3 days after inoculation (dai), no sporophores; necrotic flecks 7 dai, no sporophores; sporophores absent 3 dai but present 7 dai; sporophores present 3 dai and sporulation profuse 7 dai.

visible beyond the lesion, and 1-3 oospores are occasionally produced. Preliminary genetic analysis suggests that the response of Oy0 to CALA1 is controlled by alleles at two independent loci, whereas the response of Col0 is probably controlled by a single allele.

GENETIC ANALYSIS OF RESPONSES TO
P. parasitica

We have adopted two simultaneous approaches to the genetic analysis of responses in *A. thaliana* to *P. parasitica*. The first approach focuses on a detailed investigation of a single cross between accessions Col0 and Nd0; whereas the second approach involves an analysis of a half-diallel cross made between plants selected from fourteen accessions of *A. thaliana*.

The Col0 x Nd0 cross enables us to analyze four distinct interaction phenotypes following separate inoculations with isolates CALA2 and EMOY2 (each derived from an oospore of CALA1 and EMOY1, respectively). In response to CALA2, the fully susceptible reaction of Nd0 segregates in the cross together with the flecking phenotype of Col0. The flecking appears to be controlled by an incompletely dominant allele at a single locus (Table 3). Delayed, sparse sporulation is thought to result when the locus is in the heterozygous condition, and rarely when in the homozygous dominant condition. In response to EMOY2, the delayed sporulation of Col0 segregates in the cross together with the pitting phenotype of Nd0. All four phenotypes can be scored readily in the F_3 generation, except where delayed sporulation is masked by pitting, so we anticipate being able to determine the genotype at three loci (Table 4). The "pitting" allele from Nd0 (*RPp1*) and the "flecking" allele (*RPp2*) from Col0 are excellent candidates for molecular analysis. The allele causing delayed sporulation (*RPp4*)

Table 4. Sample of F_3 Col0 X Nd0 families scored for reaction to *P. parasitica* isolates EMOY2 and CALA2

F_3 family	Phenotypic Ratio[a]		Predicted F_2 genotype[b]		
	EMOY2	CALA2			
911448	0:0:0:10	0:0:0:12	*r1r1*	*r2r2*	*r4r4*
911455	0:2:8:0	0:12:1:0	*r1r1*	*R2R2*	*R4R4*
911449	0:3:5:2	0:3:6:3	*r1r1*	*R2r2*	*R4r4*
911430	24:2:0:0	0:0:0:10	*R1R1*	*r2r2*	*? ?*
911525	12:0:0:4	0:0:0:10	*R1r1*	*r2r2*	*r4r4*
911543	13:3:0:0	0:3:11:2	*R1r1*	*R2r2*	*R4R4*
911470	7:5:1:2	0:0:0:10	*R1r1*	*r2r2*	*R4r4*

[a] Numbers of plants in four phenotypic classes (left to right): pitting necrosis 3 days after inoculation (dai), no sporophores; necrotic flecks 7 dai, no sporophores; sporophores sparse and emergence delayed 48 h; sporophores present and profuse 3 dai.

[b] Predictions for the status of two putative loci for resistance to EMOY2. *R1* is a completely dominant allele from the Nd0 parent for pitting necrosis to EMOY2. *R2* and *R4* are incompletely dominant alleles from the Col0 parent causing flecking to CALA1 and sparse, delayed asexual sporulation to EMOY2, respectively. *R3* (not shown) is a gene from Oy0 for resistance to CALA1.

appears to be affected by factors such as the genetic background, physical environment, and inoculum concentration so fine-scale mapping will be difficult; but it should be possible to determine a coarse chromosome location.

The half-diallel analysis is an ambitious project since F_2 populations from crosses between the fourteen parents can be tested for segregation in response to numerous pathogen isolates. The research is proceeding in phases beginning with F_2 populations from crosses made between the six parents listed in Table 2. These parents were chosen because they represent accessions used commonly in laboratories worldwide. Analysis of these crosses has been nearly completed for response to two isolates (CALA2 and EMOY2). Examples from the data are shown in Table 3. Larger progeny sizes need to be tested for many of the crosses such as those where both parents are resistant to the same isolate. The remaining eight parents from the half-diallel cross include an accessions from the US (Kin0), Japan (Tsu0), Byelorussia (Ws0), and five recently collected accessions from locations in the UK where our isolates of *P. parasitica* or *A. candida* originated. Hence, it should be possible to determine whether resistance genes characterized from popular laboratory accessions can also be identified among accessions native to where the fungus can be found. In fact, preliminary F_2 data suggests that *RPp1* exists in the East Malling population of *A. thaliana* where EMOY1&2 were also collected. Complete analysis of the half-diallel is an enormous task, but we expect it to yield new alleles for investigation and thus provide a valuable resource for future investigation.

PHENOTYPIC VARIATION FOR RESPONSE TO
A. candida

Numerous parallels between the two pathosystems are expected even though two isolates of *A. candida* formed pustules on nearly all of the first sample of accessions tested (Table 1). Thirteen accessions have since been found on which sporulation does not occur or is delayed following inoculation with an isolate collected from East Malling (EMAL). Nine of these putatively resistant accessions were collected from locations in the UK and four from Germany. The responses include: no pustules and no macroscopically visible necrosis; necrotic flecking without pustules; discrete chlorotic patches without pustules; and delayed, production of minute pustules.

GENETIC ANALYSIS OF A RESPONSE TO *A. candida*

One cross is currently being analyzed in detail. A susceptible parent from accession Wein was crossed with an accession Kes37 from Keswick, UK which is resistant to the East Malling isolate of *A. candida* (EMAL). A single dominant allele for resistance to this isolate is predicted from segregation of F_2 progeny. A single allele for resistance to *P. parasitica* isolate CALA1 is also predicted; and from F_3 families that were tested for response to both EMAL and CALA1, it appears that resistance to the two pathogen isolates are controlled by alleles at different loci.

MOLECULAR MAPPING OF RESISTANCE GENES

The cross Col0 x Nd0 has already been used to construct a genetic map of Restriction Fragment Length Polymorphisms (RFLP)(9). Extensive polymorphism between the parents is therefore known and chromosome diagnostic molecular probes are available. Demonstrating the segregation of a putative resistance locus with a set of these diagnostic probes is the first step towards a fine-scale mapping of a locus. Debener *et al.* have already used the Col0 x Nd0 cross to map a locus for resistance to an isolate of *Pseudomonas syringae* pv. *maculicola* (27). The locus resides between two RFLP markers c.10 cM apart. They adopted a method referred to as "interval mapping" (71, 94) which exploits having a genetic map sufficiently saturated with RFLP markers. They also simplified their analysis by using only F_3 families from the selected class of homozygous recessive F_2 individuals, since the genotype of this class was least ambiguous.

We have begun to map the chromosome location of loci for response to *P. parasitica* using the cross Col0 *gl1* x Nd0 and two isolates EMOY2 and CALA2. Sixty-six F_3 families were produced from F_2 plants selected for susceptibility to CALA1. Another 59 F_3 families were produced from F_2 plants selected for susceptibility to EMOY2. These families are now in use for interval mapping of the respective loci for resistance to each isolate. Most of the 105 families will be useful for mapping the loci simultaneously, as illustrated in Table 4. The morphological marker of glabrous leaves (*gl1*) located on chromosome 3 was included in the cross to monitor the success of pollination. Fortuitously, the pitting allele *RPp1* maps at a locus linked close to *gl1*. *RPp2&4* appear to be at linked loci but they segregate independently from *RPp1* and *gl1*.

Concurrently, we have begun to map a locus for resistance to *A. candida* isolate EMAL using the cross Wein x Kes37 (described above). Analysis of this cross will be more complex than with Col x Nd0 because neither parent has been compared previously for genetic polymorphisms. However, initial screening of the parental genotypes with currently available probes, using six restriction enzymes, indicates that approximately 50% of the combinations tested detect polymorphisms. Polymorphisms have been detected on all five chromosomes, and it should prove feasible to map the locus for resistance to *A. candida* using the method of interval mapping. For this purpose, 43 homozygous (25 susceptible and 18 resistant) F_3 families from the cross have been produced.

CONCLUDING REMARKS

Mapping of resistance loci will initially be concentrated on alleles such as *RPp1&2*

which express clear-cut, qualitative phenotypes. The eventual objective will be to clone such genes for further molecular analysis to determine the mechanisms of resistance. However, an important challenge will be to analyze alleles that are involved in more subtle aspects of the plant-fungus interactions. With *P. parasitica*, at least, a diverse range of responses can be recognized including weakly expressed phenotypes.

Incompletely dominant or weak alleles should be included in investigations for several reasons. An example such as the phenotype characterized by delayed sporulation appears to be sensitive to experimental conditions; so it should be possible to study aspects such as genotype x environment interactions, a fundamental aspect of plant pathology. Weaker alleles will provide opportunities to study how various gene combinations affect the phenotype and could help address issues such as the effect of genetic background and apparent expression of quantitative resistance. Analysis of a range of alleles with different levels of expression may also teach us as much about fungal development as about mechanisms of disease resistance.

A curious phenomenon of "induced susceptibility" has been observed in these pathosystems and is awaiting further investigation. Accession Col0 is resistant to infection by *P. parasitica* isolate CALA1, but this accession becomes susceptible to the same isolate if it is infected previously by *A. candida* (Figure 1). The mechanism for this phenomenon could in itself be an important subject of investigation; but dual inoculations could also be useful in characterizing a class of resistance alleles. For example, alleles for resistance to *P. parasitica* could be compared for relative effectiveness in the presence of *A. candida*.

A thorough understanding of resistance in the pathosystems described hinges critically on the availability of a genetic map saturated particularly with molecular markers, and on the collection of more fungal isolates. Given a well chosen cross such as Col0 x Nd0, information can be obtained readily because molecular data from a set of F3 families or recombinant inbred lines can be used to map numerous resistance loci simply by testing the same families with different isolates of the fungus. In fact, a practical approach towards understanding the organization of alleles of Arabidopsis for resistance to *P. parasitica* will be to determine how many loci exist in each parent of the cross. Three alleles have already been found using the first two isolates collected, and we expect to find several additional resistance alleles.

We hope that this chapter has persuaded readers that fungal pathosystems of *Arabidopsis* have much to offer in providing answers to topical and important questions about interactions between plants and pathogens. We also expect that direct benefits to agricultural systems will not be long coming from such investigations.

ACKNOWLEDGEMENTS

Financial support was provided by a grant (PG207/516) from the UK Agriculture and Food Research Council as part of the *Arabidopsis* initiative of the Plant Molecular Biology programme. The second author is funded by Conselho Nacional de Desenvolvimento Cietificoe Techologico-CNPq (26.0018/91.0). We thank Dr. E. M. Meyerowitz for providing the RFLP probes necessary for mapping.

ARABIDOPSIS THALIANA AS A MODEL SYSTEM TO STUDY HOST-PATHOGENIC BACTERIAL INTERACTION

M. N. Mindrinos, G.-L. Yu, and F. M. Ausubel

Department of Molecular Biology, Massachusetts General Hospital, and Department of Genetics, Harvard Medical School, Boston MA 02114

We have developed a model system that involves infection of the small flowering plant, *Arabidopsis thaliana*, with pathogenic pseudomonads. We identified virulent and avirulent *Pseudomonas syringae* pv. *maculicola* and pv. *tomato* strains that elicit disease and resistance symptoms, respectively, when infiltrated into *Arabidopsis* leaves. We cloned an individual *P. syringae* avirulence gene (*avrRpt2*) that elicits a defense response when transferred into a virulent strain. We cloned several *Arabidopsis* defense-related genes for use in monitoring the defense response including *PAL1* (phenylalanine ammonia-lyase) and *GST1* (glutathione-S-transferase). Finally, we have isolated three *Arabidopsis* mutants that fail to give a hypersensitive response when infiltrated with a virulent *P. syringae* pv. *maculicola* strain carrying *avrRpt2*. Experiments are underway to clone and characterize the *Arabidopsis* "resistance" gene(s) defined by these mutations.

INTRODUCTION

This chapter describes the development of a readily manipulated model system to study the interaction between plants and pathogenic organisms. In the past couple of decades, significant advances have been made in studying plant defense responses at the physiological and biochemical levels and in cloning and characterizing pathogen-induced genes. Despite this effort, very little is known about the significance of particular defense responses in conferring resistance to pathogens and very little is known about the signal transduction pathways involved in the activation of these responses.

The small flowering plant *Arabidopsis thaliana* has recently become a central focus of plant geneticists and molecular biologists for reasons which have been explained above. Despite the advantages of *Arabidopsis* as a model laboratory plant, at the time we initiated our studies, there were no published reports about *Arabidopsis* pathogens. In this chapter, we describe how we have used *Arabidopsis* to study the defense response following infection by phytopathogenic *Pseudomonas* species.

Our initial goal was to identify phytopathogenic bacterial strains which elicited either compatible or incompatible responses in *Arabidopsis*. We therefore screened *Pseudomonas* strains that are known pathogens of crop plants related to *Arabidopsis* (family Crucifereae). In cooperation with Brian Staskawicz's lab at the University of California at Berkeley, we identified two virulent strains and two avirulent strains which we chose for further studies.

The second objective was to clone defense-related genes for use in monitoring their expression during the course of a compatible or incompatible interaction. This objective was accomplished using two different approaches. First, we used previously cloned defense-related genes from other plants as probes to isolate the single *Arabidopsis* gene encoding chalcone synthase (*CHS*) and one of several *Arabidopsis* genes encoding phenylalanine ammonia lyase (*PAL1*). The second approach, differential screening of a cDNA library constructed using mRNA from infected *Arabidopsis* leaves, yielded a gene encoding a glutathione-S-transferase (*GST1*), and at least two genes encoding putative calcium binding proteins.

The third objective in developing the model system was to isolate a bacterial avirulence (*avr*) gene that elicited a hypersensitive response (HR) in *Arabidopsis*. We needed a cloned *avr* gene in order to set up a genetic system in which *Arabidopsis* was responding to a signal generated by a single defined gene in the pathogen. This led to the isolation of *avrRpt2* from an avirulent strain of *Pseudomonas syringae* pv. *tomato*. When *avrRpt2* is transferred into virulent strains, it elicits a strong HR on *Arabidopsis* leaves.

The fourth objective in developing the *Arabidopsis-Pseudomonas* model system was to isolate plant mutants unable to respond to the cloned *avr* gene, *avrRpt2*. Two different screening procedures were used to identify these mutants. One method involved hand inoculation of individual plants. The other method involved the mass infiltration of plants under vacuum. To date, we have isolated three mutants which fail to respond to *avrRpt2*.

MATERIAL AND METHODS

Bacterial Growth Conditions

The *Pseudomonas* strains used in these studies were grown in King's B (KB) medium

(82) at 28°C. When present, streptomycin was at a concentration of 100 mg/L and tetra-cycline at a concentration of 10 mg/L.

Arabidopsis Growth Conditions

A collection of *Arabidopsis* ecotypes (land races) was obtained from the *Arabidopsis* information service seed bank (F. Kranz, Botanical Institute, J. W. Goethe-University, Frankfurt, Germany). Seeds were presoaked with tap water for 48 hr, incubated at 4°C for two days, plated in flats containing artificial soil (Metro-Mix 200, W. R. Grace, Cambridge MA), and covered with a plastic dome until germination (2-4 days). The seedlings were grown for two weeks in a climate controlled greenhouse (21°C ± 1°C) with supplemental fluorescent lighting (16-hr day) and then transferred to a growth chamber at 20°C with a photoperiod of 12 hr and a light intensity of 100 mE/m^2 sec.

Inoculation Procedures

Arabidopsis plants with well expanded rosettes but which had not yet bolted (4-6 weeks old) were infiltrated with mid- to late-log phase cultures (A600 nm = 0.5 O.D.) that had been centrifuged and resuspended in 10 mM $MgCl_2$. The desired number of bacteria was introduced by infiltration with a syringe, without a needle, through the underside of fully expanded leaves. Approximately 10 μl was sufficient to infiltrate half of an *Arabidopsis* leaf. To determine bacterial growth in leaves, leaf discs (0.5 cm^2) were made with a cork borer and the bacteria in the leaf tissue extracted by macerating the disc with a plastic pestle in 1 ml of 10 mM $MgCl_2$. Serial dilutions were plated on King's B plates containing appropriate antibiotics.

We have developed two mass inoculation procedures involving vacuum infiltration. In the first procedure, densely sown seeds are germinated in soil in small flats and grown through a nylon mesh. When the plants are five to six weeks old, the flats are inverted, partially submerged in a tray containing the appropriate bacterial culture, and mass infiltrated by drawing a vacuum for a few minutes in a vacuum desiccator. In the second procedure, ten-day-old *Arabidopsis* seedlings growing on petri plates are submerged in the appropriate bacterial suspension and vacuum infiltrated as above. Approximately 200 plants can be screened per plate.

RNA Isolation and Characterization

RNA was isolated by phenol-sodium dodecyl sulfate extraction and LiCl precipitation (4). RNA samples (5 µg) were separated on formaldehyde-agarose gels and transferred to GeneScreen (Du Pont-New England Nuclear, 24). The filters were prehybridized and hybridized in 0.5 M Na_2HPO_4, pH 7.2, 7% SDS and 10 mg/ml BSA at 60°C. Appropriate fragments were labeled by a random priming reaction and were added to the prehybridization buffer at a final concentration of 1×10^6 cpm/ml. The filters were hybridized 16 hr to 20 hr and then washed at 65°C for 1 hr with two changes of 2 X SSC (0.3 M NaCl, 0.03 M sodium citrate), 1% SDS.

RESULTS

Identification of *Arabidopsis* Pathogens

To identify bacterial pathogens of *Arabidopsis*, we tested *Pseudomonas* stains that are pathogens of closely related crop plants (family Crucifereae). We screened a total of 33 *P. syringae* pv. *maculicola* (*Psm*), *P.*

syringae pv. *tomato* (*Pst*), and *P. cichorii* strains for their ability to cause disease symptoms in *Arabidopsis* or to elicit a resistance response (25). The *P. syringae* pv. *maculicola* strains tested either caused disease symptoms or elicited no response, the *P. syringae* pv. *tomato* strains either caused disease or elicited a resistant response, and all of the *P. cichorii* strains tested elicited a resistant response (25,32). We chose *Psm* ES4326 and *Pst* DC3000 for further study because they were among the most virulent strains tested and because they elicited reproducible disease symptoms. At a titer of 10^5 colony forming units (cfu) per ml, *Psm* ES4326 gave disease symptoms on *Arabidopsis* leaves, which appear as water-soaked lesions with or without chlorosis 48 hr post inoculation.

Avirulent strains elicited one of the following resistance responses: no symptoms when leaves were infiltrated at 10^5 cfu/ml, hypersensitive response (HR) within 24 hr when the titer of the inoculum exceeded 10^6 cfu/ml, or chlorosis without other disease symptoms within 72 hr when the titer of the inoculum exceeded 10^6 cfu/ml. The HR in *Arabidopsis* appears as a necrotic dry lesion that develops within 24 hr depending on the strain and the inoculum used. A list of strains that have been tested in *Arabidopsis* has been published (25,32). We chose two avirulent strains, *Pst* MM1065 and *P. cichorii* 83-1, for further study because they elicited reproducible resistance responses.

Cloning an *avr* Gene

To identify and clone an *avr* gene corresponding to the resistance response elicited by *Pst* MM1065, a cosmid library of *Pst* MM1065 DNA in pLAFR3 was conjugated into the virulent strain *Psm* ES4326. We screened *Psm* ES4326 transconjugants for those that failed to elicit disease symptoms on *Arabidopsis* ecotype Columbia when infiltrated at 10^5 cfu/ml. Among 350 transconjugants tested, one transconjugant was found that did not elicit disease symptoms. The cosmid in this strain was isolated and studied further. In cooperation with Brian Staskawicz's lab, we identified a single gene, *avrRpt2*, which elicited an HR on *Arabidopsis* leaves when transferred to the virulent strain *Psm* ES4326 (32,164). Both *Psm* ES4326/*avrRpt2* and *Pst* DC3000/*avrRpt2* elicit a strong HR in most *Arabidopsis* ecotypes and display a 50-fold reduction of growth in *Arabidopsis* leaves compared to the parental strains *Psm* ES4326 and *Pst* DC3000.

To test whether other plants both related and unrelated to *Arabidopsis* could recognize *avrRpt2*, pLAFR6-76 containing *avrRpt2* was conjugated into several different *Pseudomonas syringae* pathovars, and their ability to prevent the development of disease symptoms in the appropriate host plants was tested. The most interesting result was the ability of *avrRpt2* to prevent the development of disease symptoms in various bean cultivars when present in *P. syringae* pv. *phaseolicola* (*Psp*) strain NPS3121 (123). Interestingly, when *Psp* NPS3121 was infiltrated into *Arabidopsis* leaves, it failed to multiply and did not elicit either disease symptoms or a visible HR. In contrast, *Psp* NPS3121/*avrRpt2* elicited a strong HR response in *Arabidopsis* leaves, similar to the one elicited by *Psm* ES4326/*avrRpt2*.

Cloning Defense-Related Genes

To facilitate our ability to monitor the induction of individual defense-related genes during a defense response, we used heterologous probes to clone the single *Arabidopsis* gene encoding chalcone synthase (*CHS*, 43) and one of several *Arabidopsis* genes encoding phenylalanine ammonia lyase (*PAL1*, 25). Phenylalanine ammonia lyase, the first committed step in the phenylpropanoid pathway, leads to the biosynthesis of both flavonoid and furanocoumarin phytoalexins in legumes and to the biosynthesis of lignins. Chalcone synthase is the first step in flavonoid biosynthesis. *PAL* and *CHS* genes have been shown to be induced in elicitor-treated tissue culture cells and in fungal- and bacteria-infected plants (10,17,31,30,39,66,65,93,100, 102).

We carried out a series of experiments in which RNA was isolated from leaves infiltrated with isogenic virulent or avirulent strains that differed only in the presence or absence of the single cloned *avr* gene, *avrRpt2*. We found that avirulent strains but not virulent strains strongly elicited a rapid and transient accumulation of *PAL1* mRNA in ecotype Columbia (32,25). This result suggested that *PAL1* responds directly to a signal generated by *avrRpt2*. In contrast to *PAL1*, we could not detect any increase in *CHS* mRNA levels following infection with virulent or avirulent strains (32), suggesting that flavonoid phytoalexins may not be important in defending mature *Arabidopsis* plants against pathogen attack. This result is consistent with the observation that phytoalexins, in mature *Arabidopsis* plants, are sulfonated indole compounds unrelated to flavonoids (154).

In collaboration with Brian Keith and Gerald Fink (Massachusetts Institute of Technology), we also monitored the induction of the two *Arabidopsis* 3-deoxy-D-arabino-heptulosonate 7-phosphate (DAHP) synthase genes (*DHS1* and *DHS2*, 81) which catalyze the first committed step in aromatic amino acid biosynthesis, including phenylalanine. We found that *DHS1* mRNA accumulated transiently with the same kinetics as *PAL1* mRNA. In contrast to *DHS1* mRNA, we could not detect *DHS2* mRNA in infected *Arabidopsis* leaves (81).

To identify *Arabidopsis* defense-related genes in addition to *PAL1* and *DHS1* that are activated during the defense response, a cDNA library was constructed in a pUC-derived expression vector using mRNA isolated 6 hr after infiltration of *Arabidopsis* leaves with the avirulent strain *Pst* MM1065. Eight cDNA clones that hybridized more strongly to a cDNA probe made from leaves infected with *avr* strains than to a cDNA probe made from uninfected leaves were identified and sequenced. Five of these cDNA clones encode the same protein of 208 amino acids that has 42% identity to the maize glutathione-S-transferases GSTI and GSTIII. The observation that *GST* gene(s) are induced following pathogen attack is interesting in the light of experiments carried out recently in other laboratories indicating that an important feature of the plant defense response may be a membrane-generated oxidative burst (for example, see ref. 2). A likely role for GST in an oxidative burst is the detoxification of hydroxy-alkenals, toxic byproducts of lipid peroxidation that are deleterious to membranes (1).

In addition to the *GST* gene, two cDNA clones that have homology to calmodulin genes and one clone of unknown function were also identified.

Isolation of *Arabidopsis* Defense-Related Mutants

We used *Arabidopsis* ecotypes Columbia and Nossen mutagenized with EMS or gamma rays, respectively, to isolate *Arabidopsis thaliana* mutants that do not give an HR in response to *avrRpt2*. Three different screening procedures were used. In the first procedure, a syringe without a needle was used to force a small bacterial inoculum through the stomatal openings of the underside of a leaf. We screened M2 plants for mutants that failed not only to develop hypersensitive necrosis when infiltrated with *Psm* ES4326/*avrRpt2* (16-24 hr post infiltration), but also elicited disease symptoms within 36-48 hr after infiltration. This was a laborious process, and only a limited number (several thousand) of plants could be easily screened. For this reason, only EMS mutagenized seeds, which have a high frequency of mutation (1 in 2000 for loss-of-function mutations at a single locus) were screened with this method. A single mutant was isolated among 3000 hand-inoculated plants.

The second method involved vacuum infiltration of densely sown four- to six-week-old plants growing through a nylon mesh. With this method, we were able to screen more than 35,000 M2 plants mutagenized by gamma irradiation. It was essential to screen this number because only approximately one in 25,00-30,000 M2 seeds from this seed lot carried a mutation at the *ADH* locus. A single mutant that failed to respond to *avrRpt2* was also isolated using this procedure.

The third method for isolating mutants involved vacuum infiltration of young seedlings growing on agar in petri plates. We observed that if ten-day-old *Arabidopsis* seedlings were infiltrated with *Psp* NPS3121, about 90% of the seedlings survived. However, if seedlings were infiltrated with *Psp* NPS3121/*avrRpt2*, only 5-10% survived. We interpret this result as follows: Vacuum infiltration of an entire small *Arabidopsis* seedling with *Psp*-NPS3121/*avrRpt2* elicits a <u>systemic</u> HR which usually kills the seedling. In contrast, seedlings infiltrated with *Psp* NPS3121 survive because *Psp* NPS3121 is such a weak pathogen on *Arabidopsis*. To test the method, we screened about 4000 EMS-mutagenized Columbia M2 seedlings and obtained one mutant that failed to give an HR when hand-infiltrated with *Psm* ES4326/*avrRpt2*.

To date, one gamma ray-induced mutant and one EMS-induced mutant have been characterized. Both mutants have been tested extensively in the F3 and F4 generations. In both cases, the mutation responsible for the aberrant resistant response segregates as a single Mendelian locus.

DISCUSSION

A major goal in developing an *Arabidopsis-Pseudomonas* pathogenesis system was to correlate the events leading to the development of disease resistance or the elicitation of a hypersensitive response with a single cloned avirulence gene. We achieved this goal by cloning an avirulence gene, *avrRpt2*, and then using this cloned avirulence gene to isolate *Arabidopsis* mutants that fail to prevent the development of disease symptoms and fail to give an HR in response to *avrRpt2*.

We have also cloned several *Arabidopsis* defense-related genes including *PAL1* and *GST1*. We are presently constructing transgenic *Arabidopsis* plants that carry the promoters of these defense-related genes fused to reporter genes such as *gusA* (encoding β-glucuronidase) and *luc* (encoding firefly luciferase). We plan to use these transgenic plants to isolate additional mutants that fail to activate the reporter genes in response to infection by virulent and avirulent *P. syringae* strains. These mutants should correspond to signal transduction components that act at or near the end of the signal transduction pathways. These mutants should help determine whether the activation of different defense-related *genes* involves shared signal transduction components. Finally, by obtaining mutants that fail to express specific defense-related genes, it may be possible to determine the significance of these genes in conferring resistance.

Our overall strategic approach is to isolate as _many_ different categories of *Arabidopsis* defense-related mutants as possible so that all possible genes that affect a particular phenotype being examined are identified. Analysis of these mutants should eventually enable us to determine the number of signal transduction pathways leading to defense responses, identify the components of each signal transduction pathway, and determine the significance of particular defense responses and/or specific defense-related genes for conferring resistance to a particular pathogen.

ACKNOWLEDGMENTS

This work was supported by a grant from Hoechst AG to Massachusetts General Hospital.

IDENTIFYING GENES CONTROLLING DISEASE RESISTANCE IN ARABIDOPSIS

R. W. Innes[1,2], A. F. Bent[2], B. N. Kunkel[2], and B. J. Staskawicz[2]

[1]Department of Biology, Indiana University, Bloomington, IN, 47405

[2]Department of Plant Pathology, University of California, Berkeley, CA   94720

INTRODUCTION

Our goal is to identify and characterize plant genes that are required for successful defense against plant pathogens. Because molecular cloning of such genes will likely require a map-based cloning strategy, we have chosen to investigate disease resistance in the plant *Arabidopsis thaliana* (Arabidopsis). As described in Chapter 1, this plant offers several advantages for pursuing map-based cloning projects. When we initiated this project, however, little was known about pathogens of Arabidopsis, or the general mechanisms that Arabidopsis uses to resist pathogens. In this chapter, we review our work on characterizing one bacterial pathogen of Arabidopsis, *Pseudomonas syringae* pv. *tomato* (*Pst*). We also report on our progress towards identifying Arabidopsis genes that control resistance to *Pst*.

Our first goal in developing Arabidopsis as a model for disease resistance studies was to identify bacterial avirulence genes that are "recognized" by Arabidopsis. In plants, disease resistance to specific pathogens is often characterized by a "gene-for-gene" interaction (51). In a classical gene-for-gene interaction, resistance is dependent on a single plant resistance gene that is in some way specific to a single pathogen avirulence gene; the loss of either member of this gene pair results in plant disease. Identification of a pathogen avirulence gene facilitates identification of a corresponding plant resistance gene (80).

We have previously described the isolation of a putative avirulence locus from *Pst* strain JL1065 that, when introduced into *Pst* strain DC3000, converts strain DC3000 from virulence to avirulence on Arabidopsis ecotype Col-0 (164). The *avrRpt2* clone does not affect growth of strain DC3000 in Arabidopsis ecotype Po-1, or in tomato plants. We thus believe that *avrRpt2* controls production of a product that specifically induces a resistance reaction in ecotype Col-0. To gain further insight into the mechanism of *avrRpt2* action, we have sequenced the *avrRpt2* locus. In addition, we are analyzing the genetic basis of resistance in Arabidopsis to *Pst* strains expressing *avrRpt2*. We also report here on the identification of a second avirulence gene that is recognized by Arabidopsis ecotype Col-0 and describe our preliminary work aimed at mapping a corresponding resistance gene in Arabidopsis.

MATERIALS AND METHODS

Bacterial Strains and Media

Pst strain DC3000 was obtained from D. Cuppels and *Pst* strain JL1065 was obtained from J. Lindeman. *Pst* strains were routinely cultured at 30°C on King's Medium B (82). For studies on avirulence gene expression, we used a minimal medium (13 mM potassium phosphate buffer, 17 mM sodium chloride, 30 mM ammonium sulfate, 2.8 mM magnesium sulfate, 1.7 mM sodium citrate, pH 6.8) supplemented with either citrate (10 mM, giving final concentration of 11.7 mM) or fructose (10 mM) as the primary carbon source. *E. coli* strains were grown at 37°C on Luria-Bertani medium (103). The helper plasmid pRK2013 was used in tri-parental matings (47) to mobilize broad host range vectors from *E. coli* into *P. syringae*. Matings were conducted on NYG Agar (21). Antibiotics (Sigma) were used for selection at the following concentrations: ampicillin (Ap), 100 mg/L; tetracycline (Tc), 16-20 mg/L; rifampicin (Rif), 100 mg/L; spectinomycin (Sp), 40 mg/L; streptomycin (Sm), 30 mg/L; cyclohexamide (Cy), 50 mg/L.

Plant Material, Growth Conditions, and Inoculation Procedures

Arabidopsis ecotypes were obtained from the Arabidopsis Information Service seed bank. Mutagenized populations of Arabidopsis ecotype Col-0 were generously provided by J. Ecker. Plants were grown from seed in growth chambers under an 8-hr photoperiod at 24°C. The planting and fertilizing regime was performed as described previously (164).

Plant leaves were infiltrated with *Pst* strains using a plastic transfer pipette as described (146). Two cell concentrations were commonly used, 10^6 cfu/ml and 10^7 cfu/ml in 10 mM $MgCl_2$. Cell concentrations were estimated using OD_{600}. An OD_{600} of 0.5 was found to be equivalent to about 5×10^8 cfu/ml. Following inoculation, plants were returned to growth chambers and disease phenotypes were monitored for 5 days. For inoculations at the lower cell concentration, disease level was scored on the fifth day. For inoculations at the higher cell concentration, the presence of a visible hypersensitive response (HR) was scored at approximately 24 hours after inoculation. Growth of *Pst* strains within Arabidopsis leaves was determined as described by Whalen et al.(164).

For mutant screens, a rapid dip inoculation procedure was used. Plants were grown in 4 inch pots, 10-20 per pot, in growth chambers with a light intensity of approximately 150 $\mu Em^{-2}s^{-1}$. Standard nylon screen door screening was placed over the pots at the time of sowing to hold soil in pots during the dipping process. Approximately 5 weeks after germination, plants were dipped in suspensions of bacteria (OD_{600} of 0.2) containing 10 mM $MgCl_2$ and 0.025% Silwet L-77 (Union Carbide) surfactant. Addition of the surfactant allows the bacterial solution to spread evenly over leaf surfaces. After dipping, plants were covered with a plastic dome and returned to growth chambers. Domes were removed after 24 hours and disease symptoms were scored 4 days after dipping.

Recombinant DNA Techniques and DNA Sequence Analysis.

Standard techniques were used for plasmid preparations, DNA subcloning, and agarose gel electrophoresis (4). ^{32}P-labeled DNA probes were prepared by random primer extension using the Multiprime DNA labeling system (Amersham). Southern hybridizations were performed using Hybond-N (Amersham) nylon membranes following the manufacturer's recommendations. The DNA sequence of *avrRpt2* was determined by dideoxy sequencing of single stranded DNA templates using the Sequenase 2.0 kit from U. S. Biochemical Corporation (Cleveland, Ohio).

RNA Isolation and Northern Blot Analysis.

RNA was isolated from *Pst* strains using a hot phenol extraction procedure. Cells were harvested during exponential growth, pelleted by centrifugation, and resuspended in lysis buffer (50mM Tris-HCl pH 9.0, 50 mM EDTA, 300 mM sodium acetate, 0.625% (w/v) sodium dodecyl sulfate) and incubated in a boiling water bath for 30 seconds. The lysate was then extracted 2 times with 65°C phenol and 2 times with chloroform, before precipitating with an equal volume of isopropyl alcohol. The pellet was then resuspended in deionized water and an equal volume of 4 M lithium acetate was added. The RNA was pelleted after a 60 minute incubation on ice. The pellet from the lithium acetate precipitation was then resuspended in water and the nucleic acids precipitated with ethanol. This final pellet was resuspended in water.

For Northern blot analyses, RNA samples of approximately 5 μg were run on formaldehyde-agarose gels as described by Sambrook et al. (134). Transfer of RNA to nylon membranes (Hybond-N, Amersham) was done by capillary blotting using 10X SSC (1.5 M NaCl/150mM sodium citrate, pH 7.0) as the transfer buffer. RNA was fixed to the membrane by baking at 80°C for 1 hour. Hybridizations to DNA probes and post-hybridization washes were performed following the same protocol as used for Southern hybridizations.

RESULTS AND DISCUSSION

Characterization of the Avirulence Gene
avrRpt2

We have previously localized the *avrRpt2* locus to a 1.5 kb DNA fragment (5). We have now determined the DNA sequence of this fragment (manuscript in preparation). The sequence revealed an open reading frame of 768 base-pairs. The *avrRpt2* ORF encodes a predicted protein of 256 amino acids with a molecular weight of 28.3 kilodaltons. A search of the Protein Information Resource (PIR), EMBL, and translated GenBank databases revealed no proteins with significant sequence similarities. Hydropathy analysis indicated that the AvrRpt2 protein is very hydrophilic, with no potential membrane spanning domains or leader sequences.

The sequence 5' of the avrRpt2 ORF was compared to other *Pseudomonas syringae* avirulence genes to search for conserved sequences that might indicate important regulatory motifs. We identified a 9 base-pair sequence, TGGAACCNA/C, in five out of six *P. syringae* avirulence genes for which there is sequence information available. This sequence is similar to a conserved sequence recently identified in the regulatory regions of *Pseudomonas syringae hrp* genes (46), and has also been noted by Jenner et al (76). The location of the conserved sequence relative to the first ATG of the five avirulence genes varied from -44 (*avrRpt2*) to -133 (*avrB*), however, the transcription initiation sites of these avirulence genes are unknown. This motif was not found in *Xanthomonas campestris* avirulence genes. No data is yet available relative to the function of this sequence, but it is interesting to note that many of the *hrp* genes have been shown to require the NtrA sigma factor (σ^{54}) for expression (46). In addition, expression of *avrRpt2* was found to be blocked by transposon insertions in the *hrpRS* locus of *Pst* strain DC3000 (D. Dahlbeck, C. Boucher, and B. Staskawicz, unpublished). The *hrpRS* genes have sequence similarity to *ntrC* (46), a well characterized enhancer binding protein that interacts with σ^{54} (127, 91). The conserved motif could represent a binding site for one or both of these factors.

Identification of Additional Avirulence Genes Recognized by Arabidopsis

To facilitate identification of additional Arabidopsis resistance genes, we are continuing to identify bacterial avirulence genes that induce resistant responses in Arabidopsis. We have recently found that *avrB*, a well characterized avirulence gene from *P. syringae* pv. *glycinea* (72, 143), can convert *Pst* strain DC3000 from virulence to avirulence on Arabidopsis ecotype Col-0. A third avirulence locus that is recognized by Arabidopsis ecotype Col-0, *avrRpm1*, has recently been reported by Debener et al. (27). The observation that *avrB* is recognized by Arabidopsis is especially significant because resistance in soybean corresponding to *avrB* has already been shown to be controlled by a single soybean gene, *Rpg1* (80). Because the interaction between avirulence genes and resistance genes is specific, we predict that the resistance gene in Arabidopsis that is interacting with *avrB* will be functionally and physically homologous to the genetically defined soybean gene *Rpg1*. Thus, it may be possible to clone a soybean disease resistance gene by first cloning its homologue from Arabidopsis.

Segregation Analysis of Resistance
Corresponding to *avrRpt2*

Arabidopsis ecotype Col-0 presumably contains a specific resistance gene (*Rpt2*) that interacts with *avrRpt2*. As an initial step towards cloning *Rpt2*, we have attempted to map its genetic location using a cross between ecotype Col-0 and the susceptible Arabidopsis ecotype Po-1. If Po-1 is susceptible because it lacks a single dominant resistance gene, then the ratio of resistant to susceptible F2 plants in this cross should be approximately 3:1. To our surprise, we did not obtain this result. Instead, the ratio of resistant to susceptible plants in the F2 progeny of this cross was approximately 9:7 (A. Bent, unpublished). These data indicate that resistance is controlled by more than one gene, which is contrary to a strict interpretation of the gene-for-gene model. The 9:7 ratio is consistent with a 2 gene difference, i.e. Po-1 may lack two dominant genes that are required for *Rpt2* function. If this hypothesis is true, two possible explanations are that *Rpt2* itself is a heterodimeric protein, or that Po-1 lacks a critical component in the putative signal transduction pathway between Rpt2 and the defense response. In support of the latter hypothesis, we have found that ecotype Po-1 also does not respond to two other avirulence genes, *avrB* and *avrRpm1*. The putative signal transduction pathway could be shared by multiple resistance genes.

To further investigate the interaction between Col-0 and *avrRpt2*, we have identified an additional ecotype of Arabidopsis, Wii-0, that is susceptible to Pst strain DC3000 expressing *avrRpt2* (DC3000(pAvrRpt2)). Interestingly, unlike ecotype Po-1, ecotype Wii-0 is resistant to DC3000(pAvrB). Genetic analyses of crosses between Col-0 and Wii-0, and between Wii-0 and Po-1, are currently in progress. In addition, specific F2 plants from the Col-0 x Po-1 cross have been identified that are segregating resistant and susceptible F3 progeny in an approximately 3:1 ratio. This latter result suggests that it will be possible to map, and ultimately clone, both of the putative genes in ecotype Col-0 that mediate *Rpt2* resistance.

Segregation Analysis of Resistance Corresponding to *avrB*

As described above, we are especially interested in the resistance gene in Arabidopsis corresponding to *avrB*, because it may be functionally and physically homologous to the *Rpg1* resistance gene of soybean. We are now mapping the putative *Rpg1* homologue in Arabidopsis ecotype Col-0 with the goal of cloning it via a map-based cloning strategy. We have identified several Arabidopsis ecotypes that are susceptible to strain DC3000(pAvrB). Two of these ecotypes (Bla-2 and Mt-0) have been crossed to ecotype Col-0 to generate F2 populations in which the resistance phenotype is segregating. It will be very interesting to see if resistance segregates as a single dominant gene, or as a more complicated pattern as described above for *Rpt2* in the cross between Po-1 and Col-0. Resistance in Arabidopsis to a specific strain of *P. syringae* pv. *maculicola* has recently been reported to segregate as a single gene (27). It is worth noting that the Bla-2 and Mt-0 ecotypes are resistant to DC3000(pAvrRpt2), while ecotype Po-1, which was used in the *Rpt2* mapping cross, is susceptible to both DC3000(pAvrB) and DC3000(pAvrRpt2). We are currently pursuing the possibility that Po-1 lacks a gene that is required for resistance mediated by both *Rpt2* and the putative *Rpg1* homologue.

Identification of Arabidopsis Disease Resistance Mutants

In addition to using natural variation to identify disease resistance genes and possible signal transduction genes, we are actively screening for susceptible mutants of Arabidopsis ecotype Col-0. Mutagenesis may uncover genes that are functionally conserved between ecotypes, and thus would be missed during screens of naturally occurring ecotypes.

To screen for susceptible mutants, we use a novel inoculation procedure, described in the Materials and Methods, that greatly expedites the process. This procedure involves dipping the rosettes of Arabidopsis plants in bacterial suspensions containing a small amount of surfactant. This inoculation procedure offers several advantages over vacuum or pressure infiltration methods traditionally used for bacterial pathogens. First, the method is simple and requires no special skills or equipment. Second, the bacteria infect the leaves via a "natural" route. As a result, disease susceptibility can be scored by the presence of classical bacterial speck type symptoms, e.g. small water-soaked pits with chlorotic halos, rather than entire collapse of an infiltrated region. The water-soaked pits are unmistakable, and resistant plants are easily and reliably distinguished from susceptible plants. Finally, because *Pst* is not a systemic pathogen, susceptible mutants can outgrow the infection; seed from the desired mutant can thus be obtained.

Our screen has identified two independent mutants of ecotype Col-0 that are susceptible to *Pst* strain DC3000(pAvrRpt2). The susceptible phenotype of these mutants was confirmed by monitoring bacterial growth within rosette leaves. Both of these mutants retained their resistance to *Pst* strain DC3000(pAvrB), indicating that the mutations are not in a signal transduction pathway shared by the two resistance genes. Preliminary genetic analyses indicate that these two mutants are allelic, thus only one gene has been identified to date that, when mutated, produces a clearly susceptible phenotype. We have identified several mutants, however, that may be partially susceptible to DC3000(pAvrRpt2). We are currently characterizing these mutants genetically to determine if the mutations are allelic to those described above.

SUMMARY

We are currently mapping at least two genes in Arabidopsis ecotype Col-0 that are required for resistance to *Pst* strain DC3000 expressing the *avrRpt2* avirulence gene. We have also begun to map the gene in Arabidopsis ecotype Col-0 that is required for resistance to strain DC3000 expressing *avrB*. These mapping projects represent the initial step towards molecularly cloning these genes via a map-based cloning strategy. Resistance in soybean to *Pseudomonas syringae* strains expressing *avrB* is controlled by the single dominant resistance gene *Rpg1* (80). We hope to determine whether Arabidopsis and soybean contain functionally and physically conserved disease resistance genes. We are also identifying mutants of Arabidopsis that are defective in resistance to strain DC3000 expressing either *avrB* or *avrRpt2*. We expect that this mutant screen will allow us to identify specific resistance genes, as well as genes that interact with disease resistance genes to confer the resistance phenotype.

ACKNOWLEDGMENTS

R.W.I. and A.F.B. were supported by National Science Foundation post-doctoral fellowships awarded in 1987 and 1989, respectively. B.N.K. is a DOE postdoctoral research fellow of the Life Sciences Research Foundation. This work was also supported by Department of Energy Grant DE-FG03-88ER13917 and a McKnight Foundation Individual Investigator award to B.J.S., and Public Health Service Grant R29 GM 46451 to R.W.I.

INTERACTION OF *ARABIDOPSIS THALIANA* WITH *XANTHOMONAS CAMPESTRIS*

J.E. Parker, C.E. Barber, M.J.Fan
and M.J. Daniels

The Sainsbury Laboratory, Norwich Research
Park, Colney Lane, Norwich, NR4 7UH, U.K.

Resistance to microbial diseases has been
described in many cultivated plants and has
been selected for in conventional breeding
programs. Classical genetic studies have
established gene - for - gene relationships in
several plant-pathogen systems (40,78) which
propose that a single plant product (encoded
by a resistance gene) interacts, directly or
indirectly, with a specific factor from the
pathogen (encoded by an avirulence gene)
leading to a resistant or 'incompatible'
response.

The relative simplicity of the bacterial
genome has allowed significant progress to be
made in the isolation of avirulence genes from
phytopathogenic bacteria (78,142). Their
biochemical functions are not yet understood,
although recent work shows that an avirulence
gene (*avr*D) isolated from *Pseudomonas syringae*
pathovar *tomato* acts by forming a low
molecular weight extracellular elicitor which
is then recognized by the plant(79). Also, in
several cases avirulence genes have been shown
to function beyond host-cultivar/pathogen-race
boundaries by their interaction with
particular non-host plant genotypes (78).

Most studies on plant resistance mechanisms have concentrated on changes in gene expression profiles and defence-related genes have been cloned by differential screening or by knowledge of the protein product (66,93). Such approaches are not appropriate for the isolation of plant resistance genes since these are likely to be constitutively expressed and their products are unknown. An alternative strategy is to use map-based cloning to isolate a gene defined only by its phenotype (169). In this respect *Arabidopsis thaliana* is the plant of choice since its small genome and 'rapid' genetics makes it feasible to chromosome walk to a gene of interest (113).

In our laboratory we have been studying the interaction of *A. thaliana* plants with the bacterial pathogen *Xanthomonas campestris* pathovar *campestris* (*Xcc*), the causal agent of black rot of *Brassicas* (165), and with several other crucifer-infecting *X. campestris* pathovars. We present results which show that this interaction can be considered a useful and valid model plant-pathogen system for the isolation of bacterial avirulence genes and, potentially, their corresponding plant resistance genes.

MATERIAL AND METHODS

Bacterial cultures

The laboratory isolate *Xcc* strain 8004 (21) and the mutant strains 8288 (34), 8409 (61), Tn-5C (151), 516-9 (33), ME-29 (121) and XchA2 (3), derived from *Xcc* 8004 have been described previously. *X. c.* pathovar *raphani* (*Xcr*) strain 1067 and *X. c.* pathovar *armoraciae* (*Xca*) strain 1930 were obtained from The National Collection of Plant Pathogenic Bacteria (NCPPB, Harpenden, U.K.). Spontaneous rifampicin- (rif-) resistant mutants were generated from *Xcr* 1067 and *Xca*

1930 for laboratory experiments. Cultures were maintained on nutrient (NYG) agar containing 50 μg/ml rif. All cultures containing derivatives of cosmid pIJ3200 were additionally selected on 5 μg/ml tetracycline (tet).

Growth of plants

All *A. thaliana* ecotypes were obtained from The *Arabidopsis* Information Service (89) except Landsberg *erecta* (La-er) and Columbia (Col-0) which were kindly given by C. Dean (Cambridge Laboratory, Norwich, U.K.) and JI-1 which is a local isolate. Plants were grown on a 3:1:1 mixture of John Innes No.1 compost:vermiculite:chick grit in a growth chamber at 22-24°C and 75% relative humidity with an 8 hr light period (150-200 μEinsteins/m^2/s^1). The short day light conditions promoted leaf development and delayed bolting. This was necessay to provide sufficient leaf material for inoculations and early experiments showed that the response of bolting or flowering plants to *Xanthomonas* infection was less consistent. Six to seven week old plants were taken for bacterial inoculations. Turnip (*Brassica campestris* cv. Just Right) plants were grown in the glass house at 20-25°C under a 16 hr light period. Four to five week old plants were taken for bacterial inoculations.

Inoculation of plants

Fresh overnight nutrient (NYG) broth cultures of bacteria were harvested by centrifugation, the bacteria resuspended in distilled water to an OD_{600nm} = 0.2 which is equivalent to 10^8 colony forming units per ml (cfu/ml) and 10-fold serial dilutions made from this. Bacterial suspensions were infiltrated into one half of a fully expanded leaf by gently pressing a 1ml plastic syringe without a needle to the underside. Normally four to five leaves were inoculated on each plant and the inoculated leaves were marked with a non-toxic pen.

The concentration of viable bacteria in infiltrated leaves was measured by punching out 0.2cm diameter discs from the infected area of four leaves per plant. The combined discs were homogenized in distilled water and 10-fold dilutions plated on NYG–rif agar.

Recombinant DNA techniques

Standard molecular biology protocols were used for plasmid preparations, DNA sub-cloning, ligations and agarose gel electrophoresis of DNA (103). *Xcr* 1067 genomic DNA was prepared and partaially digested with *Sau*3A to give DNA fragments enriched in the size range 20 – 30 kb as described previously (21). The DNA was cloned into the *BamH*1 site of the broad host range cosmid vector pIJ3200 (101) and introduced into *E.coli* ED8767. Recombinant clones were selected on L-agar containing tet (15 µg/ml) and then transferred 'en masse' into *Xcc* 8004 by conjugation using the helper plasmid pRK2013. Tn*5*–B20 mutagenesis and marker-exchange procedures have been described (3,155,156).

RESULTS

Infection of *A. thaliana* by *X.c. campestris*

In preliminary experiments the standard laboratory *Xcc* strain 8004 was found to cause disease symptoms on *A. thaliana* Col-0 plants similar to those incited on turnip plants (19). Several different inoculation methods, such as dipping leaves in to concentrated bacterial suspensions, infection through the leaf vein endings (33) and wound inoculations (139) resulted in disease symptom development.

However, infiltration through the stomata of the leaf underside produced the most rapid and consistent symptoms and so was used in further experiments.

Infiltration of Col-0 leaves with a suspension of 10^6 *Xcc* 8004 cfu/ml resulted in chlorosis in the inoculated area after three to four days. This spread to adjacent parts of the leaf and was followed by necrosis after five to six days (Fig. 1.A). No symptoms were observed with water or dilute buffer. The number of viable bacteria recovered from inoculated leaves increased by 100- to 1000-fold over a period of four days (Table 1, Fig. 2.). Bacteria were not recovered from uninoculated leaves up to 10 days after infiltration indicating that the infection was not systemic.

Several mutants of *Xcc* 8004 which have reduced pathogenicity (8288, Tn-5C, 516-9) or are completely non-pathogenic (ME-29, XchA2) on turnip plants were tested for their ability to cause disease on Col-0

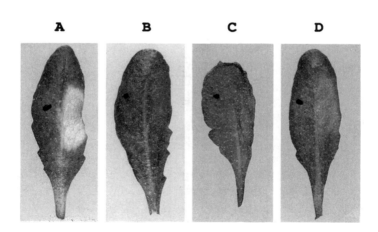

Fig.1. Symptom expression of Col-0 leaves 5 days after infiltration with *Xcc* 8004 (A) or *Xcr* 1067 (B) at 10^6 cfu/ml and 24 hr after infiltration with *Xca* 1930 (C) or *Xcc* 8004 (D) at 10^8 cfu/ml.

Table 1. Behaviour of *X.c.campestris* strains in *A. thaliana* Col-0 leaves.

Strain	Type	Symptom class	$\log_{10}$ population increase
8004	wild type	+++	2.68
8409	endoglucanase⁻ mutant	++	1.55
8288	enzyme export⁻ mutant	++	1.14
Tn-5C	regulatory⁻ mutant	++	1.24
516-9	protease⁻ mutant	+	<0
ME-29	pathogenicity⁻ mutant	–	0.03
XchA2	pathogenicity⁻ HR⁻ mutant	–	0.33

Symptoms were scored 5 days after infiltration of plants with 2×10^6 cfu/ml: –, no visible symptoms; +, mild chlorosis at inoculation site; ++, spreading chlorosis; +++, extensive chlorosis and inoculated area beginning to rot. The population increase is the factor by which the number of viable bacteria per unit area of leaf increased over a period of 4 days.

plants. All mutants were less pathogenic than wild type *Xcc* 8004 by the criterion of symptom development and growth in the leaf (Table 1). Indeed, strains ME-29 and XchA2 were completely non-pathogenic on Col-0. XchA2 is mutated in the *hrp* gene cluster, a region essential for pathogenicity on host plants and for the formation of a hypersensitive ('HR') reaction on non-host plants (3,78). The function of the gene mutated in ME-29 is not yet known (121). The results showed that the symptoms observed with *Xcc* 8004 on Col-0 are a consequence of pathogenesis and not a non-specific plant reaction to bacterial infiltration.

Natural variation in the interaction phenotype

An extensive program was undertaken to examine the interaction of 25 ecotypes of *A. thaliana* from different geographical locations, including Col-0, La-*er* and Nd-0 with 20 wild type isolates of *Xcc* to find variation in the interaction phenotype. An inoculum concentration of 10^6 cfu/ml was applied and the plants scored for symptoms for up to six days after inoculation. Results showed that there were overall differences in virulence (aggressiveness) between bacterial strains but there were no clear differential responses between ecotypes (not shown).

We therefore extended our screening to other crucifer-infecting pathovars of *Xanthomonas campestris*. *X.c. raphani* (*Xcr*) strain 1067 failed to produce symptoms on Col-0 plants at 10^6 cfu/ml (Fig. 1.B). Higher inoculum concentrations (up to 10^8 cfu/ml) also gave a null response. All 25 other ecotypes tested for their reaction to *Xcr* 1067 were symptomless. In contrast, *Xcr* 1067 produced mild disease symptoms on turnip and radish plants but it was less virulent than *Xcc* 8004 (see Table 2). Another bacterial strain *X.c. armoraciae* (*Xca*) 1930 was also symptomless on Col-0 plants and on a limited number of ecotypes tested at a concentration of 10^6 cfu/ml. However, at a high dose (10^8 cfu/ml) it incited rapid tissue collapse in the infiltrated area, characteristic of an 'HR' response (78). Confluent necrosis is observed within 16 to 24 hr (Fig. 1.C). This is quite distinct from the mild chlorosis after infiltration with 10^8 *Xcc* 8004 cfu/ml which is visible after 24 to 48 hr (Fig. 1.D).

At an inoculum concentration of 10^6 cfu/ml the titre attained by the virulent strain *Xcc* 8004 is 50- to 100-fold greater than the avirulent strains *Xcr* 1067 and *Xca* 1930 in Col-0 leaves (Fig. 2) indicating that growth of the bacteria is strongly correlated with disease symptom expression.

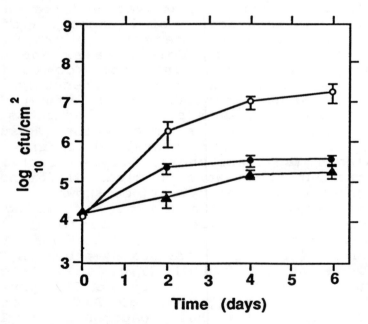

Fig. 2. Increase in *Xcc* 8004 (o), *Xcr* 1067 (▲) and *Xca* 1930 (●) in Col-0 leaves after infiltration with suspensions of 10^6 cfu/ml.

The two distinctly different resistant reactions observed with *Xcr* 1067 and *Xca* 1930 at higher doses will be analysed further to dissect the key cellular changes involved. A search is also in progress among different *A. thaliana* ecotypes for natural variation in the response to *Xca* 1930.

Cloning a putative avirulence gene
from *Xcr* 1067

In order to identify potential avirulence genes from *Xcr* 1067 a genomic library of 1067 DNA was constructed in the broad host range cosmid vector pIJ3200 (101). Clones were transferred by conjugation into the virulent strain *Xcc* 8004 and individual transconjugants infiltrated into Col-0 leaves at 10^6 cfu/ml. One clone containing a 23 kb insert (designated pIJ3130) rendered *Xcc* 8004 avirulent towards Col-0. Control experiments showed that growth of *Xcc* 8004/pIJ3130 in NYG broth and extracellular enzyme levels were the same as wild type *Xcc* 8004, indicating that general fitness of the strain was unaffected.

No symptoms were evident in Col-0 leaves inoculated with 10^6 cfu/ml *Xcc* 8004/pIJ3130 compared to 8004 containing the cosmid pIJ3200 without an insert and growth of the transconjugant was reduced to the levels of *Xcr* 1067 (Table 2). In contrast, *Xcc* 8004 /pIJ3130 was virulent to infiltrated turnip leaves, determined by disease symptom expression and bacterial growth in the leaf (Table 2). The cosmid was stable *in planta* with 8% loss over 6 days in Col-0 or turnip leaves.

The specificity of the avirulence phenotype to *A. thaliana* plants strongly suggests that the clone pIJ3130 contains an avirulence gene or genes and is not, for example, a negative-acting regulatory element which depresses pathogenicity when its copy number is increased by cloning (150).

Table 2. Growth of *Xanthomonas campestris* strains on Col-0 leaves and turnip leaves.

Strain	Col-0		Turnip	
	Symptom class	$\log_{10}$ increase	Symptom class	$\log_{10}$ increase
Xcc 8004	+++	2.94	+++	3.03
Xcc 8004/pIJ3200	++	2.51	+++	2.95
Xcc 8004/pIJ3130	−	0.85	+++	2.86
Xcr 1067	−	1.1	+	0.9

Symptoms were scored 4 days after infiltration of plants with 10^6 bacteria/ml as for Table 1. The population increase was also determined as for Table 1.

Analysis of clone pIJ3130

A combination of subcloning experiments and Tn5-B20 insertional mutagenesis has been used to identify the portion of the 23 kb insert in pIJ3130 which is responsible for avirulence activity in Col-0 plants. Insertions which caused a reversion to virulence of the transconjugant, measured by symptom development on Col-0 leaves, are clustered over approximately 2 kb. This suggests that a single gene is responsible for the avirulence phenotype which has been tentatively designated avrXca. Sequencing analysis has revealed an open reading frame of 1.8 kb which has no homology with other known sequences in the data bases. Further examination of the gene for functional motifs is in progress.

A total of 40 different *A. thaliana* ecotypes were tested for susceptibility to *Xcc* 8004/pIJ3130 and only one, Kas-1, was found to develop disease symptoms. However, growth of the transconjugant is five to 10-fold less than the wild type *Xcc* 8004. It appears, therefore, that most *A. thaliana* ecotypes recognize this avirulence gene and more ecotypes are being screened for a stronger susceptible response as a first step towards the genetic analysis of resistance to avrXca in Col-0 plants.

CONCLUSIONS

We and others (139,153) have established that the interaction of *A. thaliana* and *Xanthomonas campestris* can be utilized as a valid model plant-pathogen system. This can then be used to exploit the genetic and molecular genetic tools which are so applicable to *A. thaliana* for the isolation of genes defined only by their phenotype.

A gene was isolated from *Xcr* 1067 which renders the virulent *Xcc* strain 8004 avirulent on most *A. thaliana* ecotypes but not on turnip plants. The specificity of this effect leads us to conclude that we have identified an avirulence gene which we have designated *avr*Xca. Interestingly, this gene does not incite a hypersensitive response in *A. thaliana* plants. We are testing whether different bacterial recipient strains affect the avirulence phenotype. It is possible that the resistant reaction with *avr*Xca is a novel response quite distinct from the rapid cell necrosis characteristic of the 'HR' reported for *Pseudomonas syringae* avirulence genes on *A. thaliana* (27,32,164) but also different to the tolerance reaction observed with a wild type *Xcc* strain on Col-0 plants (153).

The use of a single defined avirulence gene within the near isogenic background of a virulent strain is proving to be a powerful strategy to identify a corresponding plant resistance gene which might otherwise be masked by the presence of multiple avirulence genes in the wild type bacterium. The ever increasing number of bacterial avirulence genes which are recognized by *A. thaliana* should allow the genetic dissection of specific and common signalling pathways in both natural and mutagenized plant populations and this is an exciting prospect in future *Arabidopsis* pathology research.

ACKNOWLEDGMENTS. The Sainsbury Laboratory is supported by the Gatsby Charitable Foundation. This work was supported in part by the Agricultural and Food Research Council. MJF thanks the Royal Society for sponsorship.

A PHYTOALEXIN FROM *ARABIDOPSIS THALIANA* AND ITS RELATIONSHIP TO OTHER PHYTOALEXINS OF CRUCIFERS

Raymond Hammerschmidt[1], Jun Tsuji[2], Michael Zook[1] and Shauna Somerville[1,2]

[1]Department of Botany and Plant Pathology
and
[2]Department of Energy Plant Research Laboratory
Michigan State University
East Lansing MI 48824

The original concept of phytoalexins as nonspecific "defensive substances" produced by a plant that was actively defending itself against fungal attack was first proposed over fifty years ago by Muller and Borger (117). This original definition was very comprehensive, and it placed many criteria on whether or not an antimicrobial compound could be truly classified as a phytoalexin. Because of the constraints of the original definition, the phytoalexin concept has undergone revisions (90, 108). Most recently, Paxton (122) proposed the following "working definition" that has been widely used for defining phytoalexins: "Phytoalexins are low molecular weight antimicrobial compounds that are both synthesized by and accumulate in plants after exposure to microorganisms." Although useful in a biochemical sense, this definition does not convey any of the important aspects of the specificity of elicitation nor does it define any role for these compounds in resistance.

This rather general definition of phytoalexins, which came about as the result of a lack of agreement on how the term should be defined (122), reflects the general lack of "hard" evidence for the relative contribution of phytoalexins in active defense. With few exceptions (e.g., the work of Van Etten and colleagues on the role of pisatin in resistance of peas to *Nectria haematococca* [152]), the evidence for phytoalexins as defensive substances comes from correlative studies on the time, rate and location of phytoalexin accumulation. To date, there has been no attempt to evaluate the role of phytoalexins through the use of phytoalexin-deficient mutants or by manipulation of the biosynthetic pathway for phytoalexin synthesis via genetic engineering.

One of our objectives is to fully evaluate the relative contribution of phytoalexins in host resistance. Our research utilizes the unique attributes of *Arabidopsis thaliana* to obtain mutants that are deficient in the biosynthesis of phytoalexins. In this paper, we will describe the general relationship of the phytoalexin from *Arabidopsis* to other cruciferous phytoalexins as well as provide information that can be used as a guide for others interested in studying this phytoalexin.

CRUCIFER PHYTOALEXINS

The first phytoalexin to be isolated and chemically characterized was the pea phytoalexin pisatin (5). Since that time, well over 150 chemically distinct phytoalexins have been characterized from at least 16 plant families (5). Phytoalexins are generally lipophilic compounds and are chemically found as sesquiterpenes, diterpenes, isoflavonoids, pterocarpans, polyacetylenes, stilbenes, flavonols and simple phenylpropanoid derivatives. If there is any common theme to the biosynthesis of phytoalexins it is that any given plant family tends to produce one chemical class of compounds as phytoalexins. For example, plants of the Solanaceae produce sesquiterpene-based phytoalexins and plants in the Fabaceae produce isoflavonoid or pterocarpan phytoalexins (5). Some of the most recently characterized phytoalexins are from the Cruciferae. These phytoalexins are all indole-based with a carbon, nitrogen and sulfur containing constituent at the 3-position of the indole ring (Figure 1).

To date, almost all of the studies on the phytoalexins of the Cruciferae have concentrated on the isolation and chemical characterization of these compounds. The 12 phytoalexins that have been characterized to date are shown in Figure 1 (6, 28, 29, 114, 115, 116, 147, 148, 153). Figure 1 also shows the structures of two stress metabolites that do not appear to function as phytoalexins (methoxybrassinens A and B, 9). In every case, these compounds are very similar in structure, with the only variation occurring in the substituent on the indole ring. This structural similarity suggests that all members of this plant family use a common indole precursor in the biosynthesis of the phytoalexins. This is similar to what is seen in the Solanaceae and Fabaceae where the phytoalexins arise from the common precursors farnesyl pyrophosphate and phenylalanine, respectively (5).

Little work has thus far been published on the role of the Cruciferous phytoalexins in disease resistance. Rouxel and co-workers have published several papers on the relative contribution of phytoalexin accumulation in species of *Brassica* (128, 129, 130). In an

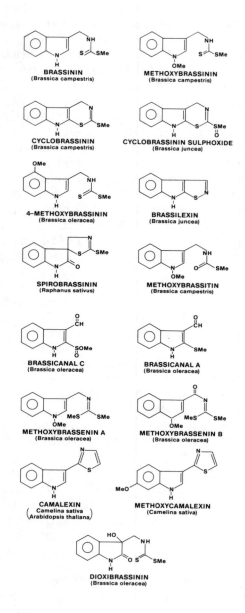

Figure 1. Phytoalexins and stress metabolites of Crucifers.

initial study, a phytoalexin later determined to be brassilexin was found to accumulate faster and to a higher level in a *Brassica* species that was resistant to *Leptosphaeria maculans* than in a susceptible species (128). Further work by this group (130), however, demonstrated that a correlation between a high level of resistance to *L. maculans* and phytoalexin accumulation occurred among *Brassica* species that contained the B genome. However, there were lines of *B. nigra* and *B. rapa* that were highly resistant but did not accumulate a high level of phytoalexins. Thus, in these interactions, phytoalexins may only be part of the defense mechanisms expressed by these plants.

Recently, two laboratories have reported phytoalexin accumulation in *A. thaliana* in response to inoculation with an incompatible pathogen (140, 154). The compound was found to possess both antibacterial and antifungal activity and was not present in the host until after infection or elicitation. Thus this compound has the characteristics of a phytoalexin as defined by Paxton (122).

Tsuji *et al.* (154) reported that the phytoalexin accumulated in leaves of *A. thaliana* after injection of the tissue with *Pseudomonas syringae* pv. *syringae*, which causes the hypersensitive response in *A. thaliana*, or by spraying with 10 mM silver nitrate. The accumulation of the phytoalexin was clearly evident by 12 hours after inoculation with *P. s.* pv. *syringae* and reached maximum levels at 48 hours. Maximum bacterial populations in the host tissues were observed at 24 hours after inoculation and then the population rapidly declined. The levels of the phytoalexin that were found in the tissue exceeded the amounts required to inhibit bacterial growth *in vitro*. Thus, the timing of phytoalexin accumulation and the decline in bacterial populations correlate well with a role for this compound in defense.

In the research reported by Tsuji *et al.* (154), the phytoalexin was purified and the structure determined to be 3-thiazol-2'-yl-indole. This compound was recently characterized as a phytoalexin of *Camelina sativa*, another crucifer, and given the trivial name "camalexin" (6, 75). Full spectral analyses and interpretation of the data supporting the structure are reported in Tsuju *et al.* (154) and Browne *et al.* (6).

PURIFICATION AND QUANTITATIVE ANALYSIS OF CAMALEXIN FROM *A. THALIANA*

Because of the ability to readily carry out genetic analysis of mutants with *Arabidopsis*, this plant may serve to be a good model to critically test the relative contribution that a phytoalexin has in host defense. However, in order to carry out this research, procedures for the analysis of the phytoalexin are needed. In this section, we describe means to isolate and quantitate camalexin from host tissue.

We have found that both inoculation by injection with *P. s.* pv. *syringae* or by spraying plants with 10 mM silver nitrate in 0.1% Tween 20 are excellent inducers of camalexin (Figure 2). Thus, analytical methods described below work equally well with either method of elicitation.

For large scale preparations, whole plants are sprayed with 10 mM silver nitrate in 0.1% Tween 20. At 24 hours after treatment, leaves are extracted with boiling 80% methanol (10 ml per g fresh wt) for 15 minutes. After cooling, the extract is filtered (Whatman #1) and the methanol removed under reduced pressure at 35°C. The remaining aqueous phase is extracted with three volumes of chloroform. The chloroform phases are pooled and the chloroform removed by evaporation under reduced pressure. The dry residue can be stored (-20°C) for later processing. For purification of camalexin, the dry residue from the chloroform extracts is dissolved in 50% ethanol and fractionated by flash chromatography on a C_{18} column (2 x 7.5 cm). The column is eluted with a step gradient of 0, 25, 50, 75, and 95% ethanol. The fractions are evaporated and then analyzed by thin layer chromatography on Silica Gel G using chloroform:methanol 9:1 (v/v). Camalexin can be detected on the chromatograms by its Rf of 0.56 and its bright blue-purple fluorescence under 302 nm UV light. Camalexin can also be identified on the plate by TLC bioassays with *Cladosporium cucumerinum*. The fractions containing the phytoalexin are pooled and the camalexin further purified via preparative TLC on silica gel using the solvent described above. Following TLC, camalexin is identified on the chromatogram by its fluorescence and then eluted from the silica gel in methanol. The methanol extracts are used for the final purification.

After removal of the methanol under a stream of nitrogen, the dry sample is purified via HPLC on a C_{18} column with a linear gradient of 1 to 100% acetonitrile. The purified camalexin is a colorless solid and has UV absorption maxima (in methanol) at 318, 275 and 215 nm (Figure 3). Camalexin has an extinction coefficient of 14,800 $M^{-1}cm^{-1}$ at 318 nm in methanol.

SMALL SCALE QUANTITATIVE ANALYSIS

For most routine analyses, smaller samples of leaf tissue are desirable. The following protocol describes a method that can readily be used for such purposes. Leaves of

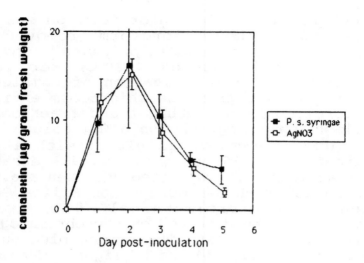

Figure 2. Time course of accumulation of camalexin in response to either AgNO3 or *P.s. syringae*.

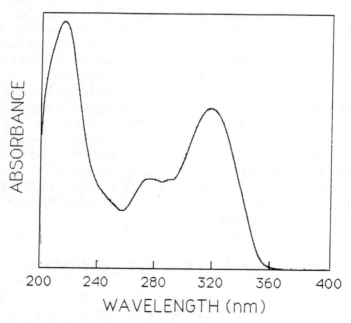

Figure 3. Ultraviolet absorption spectrum of camalexin in methanol.

three- to four-week-old A. *thaliana* plants are inoculated or treated with an eliciting agent. At the desired time, three leaves are excised from the plants. Alternately, we have found that excised leaves incubated in petri dishes on moist filter paper are also suitable for elicitation with silver nitrate. The elicited leaves are placed into 7 mls of 80% methanol in a test tube. The tubes are heated at 85°C until the volume of the extract is reduced to about 2 ml. The solution is then decanted into a clean text tube and the volume adjusted to 3 ml with 20% methanol. This solution is then applied to a 3-ml C_{18} solid phase extraction column (Supelco) previously conditioned with 2 ml of methanol followed by 2 ml of water (care must be taken to never allow the column to run dry). We have used an elution rate of 2 ml per minute using a Visiprep vacuum manifold (Supelco). After application of the sample to the conditioned column, the column is washed with 2 ml of 30% methanol. Camalexin is then eluted from the column with two 1.5 ml portions of 80% methanol. The extract can then be analyzed for camalexin content by several means.

FLUOROMETRIC QUANTITATION

Camalexin is an intensely fluorescent compound, and we have found that it can be readily quantified by fluorometry. The 80% methanol eluates from the C_{18} extraction columns can be used directly for fluorometric quantification of camalexin. The excitation wavelength of the fluorimeter is set at 330 nm and the emission wavelength set at 393 nm. A typical fluorometric standard curve of camalexin in 100% and 70% methanol is shown in Figure 4 (note that there is an effect of the water content on the standard curve). In addition to the direct fluorometric measurements, we routinely check the eluates from the C_{18} column by TLC to confirm the presence of camalexin. The sensitivity of this method is suitable for most routine analyses, and it is sensitive enough to allow analyses on samples smaller than three leaves.

UV ABSORPTION QUANTITATION

The amount of camalexin can also be determined via UV spectroscopy; however, some additional steps are needed. The extracts from the C_{18} extraction columns can be used after evaporation of the solvent and redissolving the residue in a small amount of methanol. The methanol extracts are then used to separate the camalexin via thin layer chromatography as described below. Alternately, we have found that the 80% methanol extracts obtained in the small scale extraction procedure can be used without the C_{18} step. In brief, after gentle heating of the 80% methanol extracts to remove the MeOH, H_2O is added to the original extraction volume. The solution is then extracted twice with $CHCl_3$ (v/v) and the $CHCl_3$ extracts are combined. After removal of the chloroform under reduced pressure, the residue is dissolved in a small amount of chloroform and applied to a silica gel G TLC plate. After development in chloroform:methanol (9:1, v/v), the camalexin is located under UV light. The area of silica gel containing the camalexin is removed and eluted with 1 ml of spectral grade methanol. The absorption of the methanol containing camalexin is determined at 318 nm and the amount of camalexin calculated from its extinction coefficient. A standard curve of pure camalexin is shown in Figure 5.

In all quantitation procedures described here, no adjustments were made for losses of the compound during extraction and

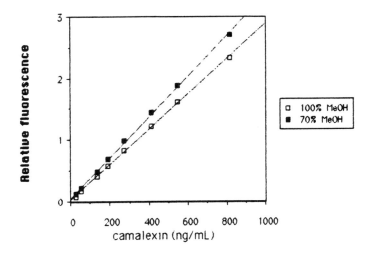

Figure 4. Fluorescence emission standard curve for pure camalexin in 70% or 100% methanol. Determination made with an SLM Aminco Fluorometer.

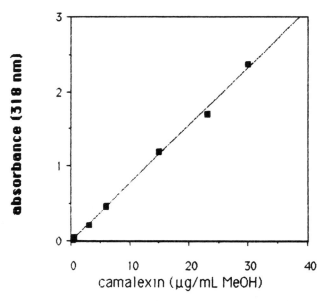

Figure 5. Ultraviolet absorption standard curve for pure camalexin in methanol.

purification prior to the analyses.

CONCLUSIONS

We have demonstrated that *Arabidopsis thaliana* has the ability to synthesize what appears to be a typical phytoalexin. The relative ease of genetic manipulation and mutational analysis of *Arabidopsis* suggests that this host will prove valuable in studies of the role and relative contribution of camalexin in disease resistance responses that are currently being described in several *Arabidopsis*-pathogen interactions. The same genetic attributes of *Arabidopsis* will also help facilitate future studies on the biochemical pathways that lead to camalexin synthesis, how this pathway is regulated, and whether camalexin synthesis is coordinately regulated with other putative defense responses in *Arabidopsis*.

ACKNOWLEDGMENTS

We gratefully acknowledge the financial support of the Michigan Agricultural Experiment Station, the U.S. Dept. of Agriculture Competitive Grants Office and U.S. Dept. of Energy (DE-FG02-90ER20021).

ARABIDOPSIS THALIANA AND TURNIP CRINKLE VIRUS: A MODEL PLANT-PATHOGEN SYSTEM

A. E. Simon[1], M. Polacco[2], X. H. Li[3], J. E. Lew[4], R. E. Stange[1], and C. D. Carpenter[1]

[1]Department of Biochemistry
and Molecular Biology
University of Massachusetts, Amherst MA 01003

[2]Department of Biochemistry
University of Missouri, Columbia MO 65211

[3]Department of Plant Pathology
University of Massachusetts, Amherst MA 01003

[4]Department of Botany
University of Massachusetts, Amherst MA 01003

The complexity of the plant response to both virulent and avirulent pathogens presents a major difficulty in elucidating the molecular events involved. Although the genetics underlying compatible and incompatible interactions between plants and pathogens has been defined for a number of systems (52, 54), little is understood about these interactions at the molecular level.

We have begun the process of developing a new system for the study of plant-virus interactions using *Arabidopsis thaliana* as a host. The virus we have chosen is turnip crinkle virus (TCV), a member of the carmovirus group. TCV is one of the smallest and simplest of the plant RNA viruses with a

single stranded genome of 4054 bases (8, 11).
The five open reading frames specify the viral
capsid protein, replicase functions and
proteins required for systemic movement in
plants.   TCV is also one of the best
characterized viruses at the structural level
(7, 163).

TCV is unusual in its association with a
number of small, subviral RNAs.   These RNAs
are either derived from the viral genomic RNA
(defective interfering RNAs; 97), unrelated to
the viral genome (satellite (sat-) RNAs; 138),
or recombinant molecules derived from segments
of both the viral genome and a sat-RNA (138,
170).   Subviral RNAs which include a segment
from the 3' ends of the TCV genomic RNA
exacerbate symptoms on a variety of hosts
(98).   In this chapter, we will describe
studies which have established the TCV-*A.
thaliana* system and report preliminary results
on some of the molecular events which are
triggered by infection of a susceptible
ecotype with TCV.

MATERIALS AND METHODS

Virus Isolates

Two isolates of TCV have been described.
The TCV-B isolate has been studied in the lab
of T. J. Morris (University of Nebraska) and
is characterized by the production of mild
symptoms on the host regardless of the
presence of DI or sat-RNAs (11).   TCV-M (also
known as TCV-JI), the isolate routinely used
in our laboratory, produces highly virulent
symptoms on a variety of hosts when associated
with the hybrid DI/sat-RNA named sat-RNA C.
Both the TCV-B and TCV-M have been sequenced
(8, 11, Carpenter, Song and Simon, unpublished
results),   and   inoculation   of   *in vitro*
synthesized full length transcripts produces a
systemic infection on host plants (68;

Carpenter, Song and Simon, unpublished results). We routinely propagate the virus in turnip cv Just Right. Work with TCV is strictly regulated by the USDA since this virus is not presently found in the United States.

A. *thaliana* Ecotypes and Plant Growth Conditions

A. *thaliana* ecotypes used in our work were the generous gifts of S. Somerville (Michigan State University) and F. Ausubel (Harvard Medical School). Unless otherwise noted, A. *thaliana* seeds were sown in Promix (Premier Brands, Inc.), then germinated and grown in a growth chamber at 20°C, 15,000 lux with a day length of 14 hours. Vernilizing the seeds for five days did not make any appreciable difference in the number or uniformity of the germinating seedlings.

Inoculation of A. *thaliana*

Unless otherwise noted, A. *thaliana* seedlings with four fully expanded leaves (approximately 14 days after sowing the seeds) were dusted with celite and then inoculated by rubbing each leaf with one stroke of a glass rod dipped in infection buffer (0.05 M glycine; 0.03 M K_2HPO_4; 0.02% bentonite, pH 9.2) containing 0.1 mg/ml of total plant RNA extracted from turnip infected with TCV-M. Our studies have shown that inoculating only one or two leaves is not sufficient to achieve 100% infected plants.

Detection of Virus in TCV-Infected Plants

In general, viruses can be detected in infected tissue by several methods: electron microscopy; ELISA assays using specific anti-coat protein antibodies; extraction and analysis of plant nucleic acids; *in situ* hybridization to sectioned tissues; *in situ* hybridization to the whole plant (whole plant blots). The two methods we use to detect TCV in infected *A. thaliana* were chosen based on the prevalence of the virus in infected tissue. TCV genomic RNA accumulates in vast quantities in infected plants, approaching the level of cytosolic ribosomal RNAs. The subviral RNAs are also amplified very efficiently, becoming some of the most prevalent RNAs in the cell. This level of accumulation of TCV genomic and subgenomic RNAs permits the simple detection by electrophoresis using 1.2% non-denaturing agarose gels stained with ethidium bromide. RNA is first denatured by heating for 10 min at 70°C in 100% formamide plus bromphenol blue. For quantifying the level of accumulation of viral RNAs, the gels are treated with 6% formaldehyde for 20 min and the RNA transferred to Nitran membranes (Schleicher and Schuell) and hybridized to radiolabeled probe using standard protocols (4).

The second technique we use for detecting the virus is the whole plant blot, originally developed for use with DNA viruses (110). We have modified this technique for use with RNA viruses. Briefly, plants are soaked in ethanol and then treated with SDS and pronase followed by HCl. Plants are then subjected to hybridization using standard conditions. This technique has been used to monitor virus movement and levels in resistant and susceptible *A. thaliana* ecotypes (Simon *et al.*, in press).

Viral RNA Isolation

We have developed a quick and efficient method for isolating single-stranded RNA from infected plants. Plants (which can be kept at

-80˚C until ready for use) are ground in liquid nitrogen in a 50 ml beaker using a small spatula or pestle. Ground material is then transferred to a 1.5 ml eppendorf tube up to a volume of 0.5 ml. RNA extraction buffer (0.2 M Tris-HCl, pH 9.0; 0.4 M LiCl; 25 mM EDTA; 1% SDS) 0.55 ml is added and the tube vortexed vigorously. H_2O-saturated phenol 0.55 ml is added and the tube centrifuged for 2 min. The aqueous layer is re-extracted once with phenol, then once with chloroform followed by ethanol precipitation of the nucleic acids. The pellet is resuspended in 0.3 ml of ice cold 2 M LiCl and then centrifuged at 4˚C. The pellet is resuspended in 0.3 ml of H_2O followed by a final ethanol precipitation. One person can prepare over 30 RNA isolations in a single day. Omitting LiCl from the extraction buffer and the LiCl precipitation step results in the isolation of total nucleic acids including double stranded RNA forms. However, we have found that this omission results in a high amount of contaminating polysaccharides which can affect the migration of RNA in gels.

RESULTS

Susceptible and Resistant Ecotypes of
A. thaliana

In order for the TCV-A. thaliana system to be a useful model of plant-virus interactions, it is necessary to identify both resistant and susceptible A. thaliana eco-types. Towards this goal, we initially tested six ecotypes of A. thaliana for susceptibility to the virus: La-O, Col-O, Ag-O, Bus-O, and No-O. All were found to be equally susceptible to TCV infection if inoculated when plants had four fully emerged leaves (98). All infected plants ceased growing by six days following inoculating. At this time, leaves began to curl under, then became increasingly necrotic over the next two weeks by which time all infected plants were dead. The lethal nature of the virus was only manifested when TCV was associated with the subviral sat-RNA C. Inoculating plants with virus cured of sat-RNA C resulted in stunted, chlorotic plants which were still able to bolt and set seed. A. *thaliana* plants inoculated when plants had four fully emerged and four emerging leaves (about 21 days post-sowing) differed in their susceptibility to the virus. Especially noteworthy was the La-O ecotype which was no longer killed by the virus three weeks following inoculation. However, these plants still displayed severe symptoms including chlorotic leaves and stunting due to short internode length.

The prevalence of TCV genomic and subviral RNAs in infected tissue simplifies the detection of viral nucleic acids. Total single stranded plant RNA is subjected to electrophoresis on 4% denaturing acrylamide gels and then stained with ethidium bromide. An example of such a gel is shown in Fig. 1. In this experiment, Col-O seedlings were inoculated with viral RNAs and total single stranded plant RNA was prepared one, two or three days later. The accumulation of the sat-RNAs C and D was evident after only two days of infection. It is this rapid accumulation of sat-RNAs (including the virulent sat-RNA C) which is probably responsible for the severe effect of the virus on A. *thaliana*.

Because of the widespread use of the ecotype Col-O by groups studying A. *thaliana*, we use this ecotype as our primary susceptible host. We approached finding a resistant line of A. *thaliana* in two ways. First, in an effort to identify a resistant, isogenic line of Col-O, we inoculated almost 6500 seedlings derived from selfing progeny of ethyl-methyl-sulfanate (EMS) treated seeds (gift of F.

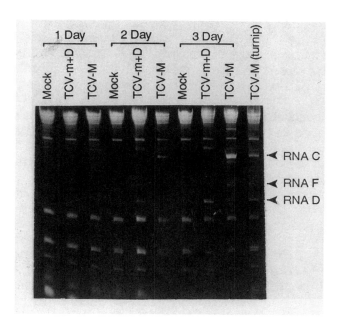

Fig. 1. Accumulation of TCV subviral RNAs in the susceptible ecotype Col-O. Seedlings were treated with buffer only (mock), with TCV genomic RNA plus sat-RNA D (TCV-m+D), or with TCV genomic RNA and the three sat-RNAs C, D, and F (TCV-M). Total plant single stranded RNA was subjected to electrophoresis on a 4% denaturing polyacrylamide gel which was subsequently stained with ethidium bromide.

Ausubel). This approach proved unsuccessful as all infected plants died. A second approach involved testing 20 additional ecotypes. Of these, only the ecotype Dijon (Di-O) showed excellent resistance to the virus. Virus-inoculated Dijon exhibited few, if any, symptoms early in the infection; however, in some experiments, 10-15% of Dijon plants which had bolted and set seed exhibited mild symptoms such as bolt curling and early desiccation. Northern hybridization analysis has indicated that viral genomic RNA is detectable in Dijon at eleven days post-inoculation. Resistance to TCV by Dijon is probably associated with this lack of early virus replication and/or movement.

Effect of Environment on Susceptibility and Resistance of A. *thaliana* to TCV

In order to ascertain the impact of environmental parameters on the ability of TCV to infect A. *thaliana*, we have monitored the effect of day length and light intensity on virus-infected plants. These two parameters are known to affect other plant-virus systems (107). For the day length experiments, we grew Col-O and Dijon in 16,000 lux for eight hours/day up to 24 hours/day. For the light intensity experiments, infected plants were subjected to 16 hours/day of light from very low intensity (10,000 lux) to very high intensity (30,000 lux).

Preliminary results indicate a distinct pattern: conditions which cause the plants to grow most rapidly (30,000 lux or 24 hour day length) also result in the quickest death for the susceptible Col-O ecotype and the most consistent resistance by the Dijon ecotype. Lower light intensity and/or decreased day length result in variability among the individual plants. At the lowest light intensity or day length, not all Col-O plants inoculated with the virus become infected which differs considerably from the virtual 100% infection rate of Col-O grown at 15,000 lux, 16 hours/day.

Genetics of Resistance to TCV

One of the advantages of using A. *thaliana* is the short generation time (about six weeks). A. *thaliana* produces very small flowers and crosses are generally done with

92

the aid of a low power dissecting microscope. Reciprocal crosses between Dijon and Col-O result in heterozygous progeny which, when infected with TCV, exhibit a delay in symptom manifestation of about two days. The infected F1 plants eventually become completely necrotic like the susceptible parent. Selfing of the F1 plants produced an F2 generation, and preliminary results suggest that the resistance phenotype is segregating in a 1:2:1 ratio typical of a single copy gene. Recombinant inbred lines are currently being generated and future experiments will involve the cloning of the resistance gene. *A. thaliana* will be particularly useful for this aspect of the project because of the excellent chromosome maps available and small genome size.

Plant Responses in Susceptible Interactions

Differences in proteins accumulating in response to viral infection between Dijon and Col-O ecotypes is apparent a few days post-inoculation. Dijon and Col-O proteins isolated from plants one day after inoculation with either virus or buffer and separated on two-dimensional gels under our conditions are virtually indistinguishable. At three days post-inoculation, non-TCV-encoded proteins begin to accumulate in infected Col-O which are not present in mock (buffer) treated plants. The induced proteins (and others) become much more prevalent at six days post-inoculation, the time when symptoms are first visible (Fig. 2). Virus-induced proteins were not discernible in Dijon plants at six days post-inoculation.

To characterize proteins which are induced in response to virus infection, a genomic library of *A. thaliana* DNA, obtained from Dr. Elliot Meyerowitz, was screened using cDNA populations derived from RNA isolated

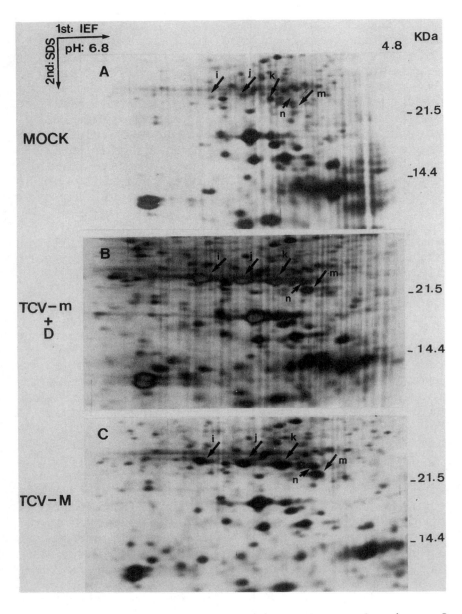

Fig. 2. Two-dimensional gel analysis of proteins accumulating in Col-O plants in response to virus infection.

from four-day virus-infected or mock-treated plants. We had previously shown that mRNA populations accumulating four days after virus inoculation of *A. thaliana* specified *in vitro* translation products which were not found in mock-treated plants. This strategy has so far produced two groups of overlapping phage which contain genomic sequences coding for virus-induced proteins. We have completed the sequencing of both the genomic clones and the corresponding cDNAs which were isolated from a cDNA library prepared from mRNA isolated four days after infecting plants with TCV. Our results indicate that both clones specify novel glycine-rich proteins. One of the proteins shares substantial identity (over 77%) with glycine-rich proteins from maize and sorghum (13, 60). One intriguing property of this latter class of glycine-rich proteins (as opposed to the more common glycine-rich proteins which are presumably associated with the plant cell wall) is that they belong to a class of proteins, mainly studied in animal systems, which bind to specific RNAs and contain a characteristic RNA-binding sequence (36). In maize, transcription of this glycine-rich protein is induced by water stress or abscisic acid and the protein does not contain a signal sequence (60).

DISCUSSION

In this chapter, we describe a new system for studying resistance and susceptibility to plant RNA viruses. The advantages of using *A. thaliana* in the study of plant-pathogen interactions have been discussed in the introduction to this symposium book and need no repetition here. TCV is a particularly good virus to use in a model system because of the small size of the single TCV genomic RNA, the ease of inoculating by mechanical methods, the existence of both mild and severe isolates, the infectivity of *in vitro* synthesized transcripts, the intriguing association with numerous small subviral RNAs and, most importantly, the ability to cause a lethal infection on most ecotypes of *A. thaliana*. All of these attributes enhance the potential of TCV to provide new levels of understanding of basic viral processes.

As part of a multifaceted approach to the study of the interaction between a virus and its host, we have begun to characterize both resistant and susceptible responses to TCV by *A. thaliana*. Susceptible and resistant ecotypes have been identified; the susceptible interaction is characterized by rapid symptom progression leading to the death of the plant by 21 days post-inoculation. A variety of virus-induced proteins accumulate by six days following inoculation and we have begun to characterize these proteins at the molecular level. The resistance response by Dijon is probably due to a lack of virus accumulation early in the infection process. Preliminary results on the genetics of resistance indicate that resistance is probably cause by a single copy, partially dominant gene. Initial studies on the effects of light and day length demonstrate the influence of the environment on the interaction between plant and pathogen.

CONCLUSION

In summary, many unanswered questions remain in our understanding of plant-virus interactions. What does resistance mean in molecular terms? How do viruses elicit disease symptoms? What host factors are involved in basic virus processes such as replication and movement? Despite a prodigious amount of effort by virologists throughout the world, progress has been slow, and breakthroughs in our understanding have been few and far between. It is hoped that the development of simple model systems such as TCV-*A. thaliana* will provide answers to some of these fundamental questions.

ACKNOWLEDGMENTS

This work was supported by National Science Foundation grants DMB 9004665 and DMB 9105890 to A. E. S.

ARABIDOPSIS AS AN EXPERIMENTAL HOST PLANT OF THE PHYTOPATHOGENIC MOLLICUTES

J. Fletcher, D. A. Golino, and C. E. Eastman

Department of Plant Pathology, Oklahoma State
University, Stillwater, OK 74078
and
USDA-ARS, Department of Plant Pathology,
University of California, Davis, CA 95616
and
Center for Economic Entomology, Illinois
Natural History Survey, Champaign, IL 61820

Arabidopsis thaliana (L.) Heynh. has been used for studies of numerous phytopathogens including viruses, fungi, and bacteria. Recently, this versatile plant has been found to be susceptible to the Mollicutes, a group of cell wall-less prokaryotes including the cultivable spiroplasmas and the as-yet uncultured mycoplasmalike organisms (MLOs).

The mollicutes include pathogens that are found worldwide and cause disease in a broad range of annual and perennial plants (109). However, technical difficulties hamper efforts to characterize these pathogens (83). Those phytopathogenic mollicutes that have been identified are transmitted in nature only by phloem-feeding leafhoppers (104). Lacking a cell wall and confined to the plant phloem, they are fragile and difficult to purify from both plant and insect hosts. Although several plant pathogenic spiroplasmas have been cultured, no MLOs have been cultured. Also, no efficient transformation or transfection system is available for these prokaryotes and codon usage in mollicutes is unique (UGA coding for tryptophan), making expression of their genes in *E. coli* difficult. These problems limit the study of pathogen strains and/or mutants.

Since the relationship of the leafhopper vectors with mollicutes is highly specific, the natural plant host ranges of these pathogens are heavily influenced by feeding preferences of their vectors.

Many phytopathogenic mollicutes infect the Brassicaceae. *S. citri* affects several species (48, 120) and McCoy et al. (109) listed thirty-one naturally-occurring plant diseases caused by MLOs in brassicas. Although it is unlikely that the causal agents of all these diseases are distinct MLOs, clearly the mollicutes are widespread among both cultivated and non-cultivated brassicas.

Two mollicutes have been demonstrated to infect *A. thaliana* (50, 57). *Spiroplasma citri* is the helical, cultivable causal agent of citrus stubborn (20) and horseradish brittle root (48). The beet leafhopper transmitted virescence agent (BLTVA) is a non-cultivable MLO found naturally in several brassicas (58). Both of these agents are transmitted by the beet leafhopper, *Circulifer tenellus*.

A. thaliana may be a valuable experimental host of mollicutes. Many of the symptoms they cause, including chlorosis, stunting, virescence, phyllody, premature flowering, and the proliferation of floral and vegetative organs, suggest that mollicutes disrupt the host's normal metabolic processes in unusual ways. Understanding these changes in host physiology could be facilitated by the use of *Arabidopsis* as a system for the study of plant-mollicute interactions.

MOLLICUTE SOURCES AND MAINTENANCE

S. citri. *S. citri* strain BR3, originally cultivated from Illinois horseradish affected by brittle root disease (48), was maintained in a series of leafhopper-inoculated turnip plants (*Brassica rapa* L.) in a growth chamber (27:22 C, 16L:8D). Spiroplasmas were cultured in LD8 broth at 31 C (48).

BLTVA. Since the MLOs have not yet been cultured axenically, the type BLTVA line FC-83-13 (55) used for these studies was maintained in either leafhopper vectors or plants. Infected daikon radish, *Raphinus sativus* L. 'Summer cross hybrid', and periwinkle, *Catharanthus roseus* (L.) Don. 'Little Pinkie,' were used as source hosts.

VECTOR SOURCES AND MAINTENANCE

S. citri. A colony of healthy *C. tenellus* (originally collected in 1979 from Illinois horseradish fields) provided insects for transmission of *S. citri* to turnip and *Arabidopsis* plants. Mid-sized nymphs were given a 7-day acquisition access period on infected turnip, held 14 days on healthy sugar beet (*Beta vulgaris* L.) (a nonhost of *S. citri*) and then confined as adults on test plants. All phases of transmission were conducted in a growth chamber (27:22 C, 16L, 8D).

BLTVA. BLTVA was transmitted using a colony of *C. tenellus* reared on *B. vulgaris* 'VH510.' Healthy colonies were checked regularly by feeding on indicator periwinkle to insure that the colonies were free of leafhopper-borne plant diseases. Colonies were maintained in a glass house (30:20 C,

16L, 8D). BLTVA-inoculative leafhoppers were produced by rearing the insects on infected daikon radish, in growth chambers (27:22 C 16L, 8D).

PLANT SOURCES AND MAINTENANCE

S. citri. *Arabidopsis* ecotypes Lansburg Erecta, Columbia, and Niedersans (initial seeds obtained from Dr. David Meinke, Oklahoma State University) were greenhouse-grown in a soil-peat mixture in clay pots (7.6 cm) placed on trays containing charcoal.

BLTVA. Wild type *A. thaliana*, ecotype La-o, and the ga-1 mutant were provided by Dr. M. Koornneef (Agricultural University, Wageningen, The Netherlands). Seed was sown in a suspension of agar upon a light, well-drained commercial potting mix. Flats of seedlings were grown in growth chambers under short days to encourage rosette growth and delay flowering for inoculation experiments (25:20 C, 8L, 16D).

INOCULATION OF PLANTS WITH MOLLICUTES

S. citri. One to several inoculative leafhoppers were confined on young plants (one or more plants per cage) in the growth chamber. Plants caged without leafhoppers or with non-inoculative insects served as controls. After the inoculation access period, leafhoppers were removed and the plants were held in the growth chamber.

Single leafhoppers were able to transmit *S. citri*, although transmission rates increased with additional leafhoppers. However, prolonged feeding by many insects, healthy or infected, on individual seedlings was deleterious to the small plants. The overall infection rate of *Arabidopsis* with *S. citri* was 43/51 exposed plants (84%); with ecotype Lansburg at 84%, Columbia at 80%, and

Niedersans at 90%. *C. tenellus* was observed to feed readily and oviposit on *Arabidopsis*. Nymphs were produced on healthy and *S. citri*-infected plants. Although the nymphs were not allowed to develop into adults on these plants, it is likely that *Arabidopsis* is a suitable feeding and breeding host for *C. tenellus*. Wild and cultivated brassicaceous plants are favored hosts for this leafhopper (12).

BLTVA. In the initial experiment to determine susceptibility of *Arabidopsis* to BLTVA, 16 individual plants were caged with 5 inoculative adult leafhoppers per plant. The same number of non-inoculated control plants were caged. After one week of leafhopper feeding, normally more than sufficient to transmit BLTVA, all plants were fumigated with DDVP (2,2-diclorovinyl dimethyl phosphate) to eliminate the vectors and their eggs. Plants were observed for symptom development for several weeks. Eighty-seven percent of the inoculated plants became infected, which is typical for most hosts. The ga-1 mutant was also readily infected.

DETECTION OF THE PATHOGEN IN INFECTED PLANTS

Both *S. citri* and BLTVA were readily detected in infected *Arabidopsis* using techniques normally used in mollicute research in our laboratories.

Symptomatology

S. citri. Symptoms of *S. citri* infection, which began to appear 14 days after leafhoppers were placed onto plants, were similar among the three *Arabidopsis* ecotypes. Infected plants often had small or absent rosettes, stunted floral stalks (Fig. 1), and curled or deformed cauline leaves (Fig. 2). Internodes on flower stalks were shortened,

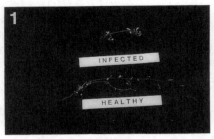

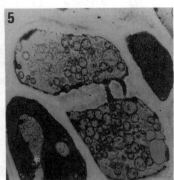

Figure 1. *Spiroplasma citri*-infected (top) and healthy *Arabidopsis thaliana*. Note stunting of the floral stalk, reduced silique size and internode length, and terminal bunching of siliques and flowers. Infected plant 7.2 cm long, healthy plant 23.5 cm long.

Figure 2. *S. citri*-infected *A. thaliana* plant showing curled and C-shaped cauline leaves. Note normal flowers.

Figure 3. *S. citri*-infected *A. thaliana* plant showing necrotic siliques and terminal bunching of floral organs.

Figure 4. *A. thaliana* ecotype La-o, healthy (left) and infected with the beet leafhopper transmitted virescence agent (right). Note the phyllody, virescence and proliferation of floral organs.

Figure 5. Transmission election micrograph showing phloem of *A. thaliana*, infected with BLTVA, filled with the characteristic membrane bound cells of this pathogen (photo by M.E. Shaw).

with terminal bunching of flowers and siliques (Fig. 3). Flowers and siliques were sometimes reduced or necrotic, although the pedicel remained green. Seed set was often significantly reduced. These symptoms are similar to those of *S. citri* in other brassicaceous plants (48). No virescence, phyllody, or asymmetry of floral structures was observed. Infected plants declined and died more quickly than healthy controls.

BLTVA. Four weeks post inoculation, both infected and control La-0 plants produced white flowers, followed by full siliques and seed. In the fifth week, virescence and phyllody symptoms characteristic of BLTVA developed (Fig. 4). No subsequent flowers were normal. Proliferation of secondary flowers was also observed; the pistil of infected flowers enlarged and grew as a stem producing virescent flowers. Stem elongation within these distorted flowers was noted. As is typical of BLTVA infection of brassicas, foliar symptoms were not observed. In contrast to *S. citri* infected plants, leaves remained green and of normal size.

Because BLTVA infection caused abnormal, sterile flowers, the infected plants did not set seed and, therefore, continued to grow vigorously after the healthy plants had become senescent. Longer days might reduce the time required for flowering and allow earlier observation of symptoms.

Cultivation in artificial medium and PAGE

Spiroplasmas were cultured from *S. citri*-exposed *Arabidopsis* plants into LD8 broth (48). Positives included symptomatic plants and occasional asymptomatic plants exposed to spiroplasma-infected leafhoppers, but no control plants. PAGE protein profiles (48) of spiroplasmas cultured from *Arabidopsis* were indistinguishable from those of characterized BR3. None of the plant pathogenic MLOs can yet be cultured in artificial medium (109).

Enzyme linked immunosorbent assay (ELISA)

S. citri and BLTVA both could be detected by ELISA using antiserum for the respective pathogen. Foliage or flowering stalks of *Arabidopsis* infected with *S. citri* were ground (2/1) in buffer for double sandwich ELISA (49). Absorbance readings at 405 nm for healthy *Arabidopsis* samples were indistinguishable from buffer readings. Readings for infected samples varied within expected ranges. Likewise, detection in BLTVA-symptomatic plants was possible using a F(ab')2 ELISA system (59). No significant differences were noted between the values of BLTVA-infected tissue of *Arabidopsis* or of other hosts.

Electron microscopy

Arabidopsis tissue with BLTVA symptoms was harvested 7 wk post inoculation, fixed, dehydrated in a standard ethanol series and imbedded in Spurr's medium (Polysciences Inc., Warrington, PA 18976). When phloem sections were examined by transmission electron microscopy, typical mycoplasma-like bodies (Fig. 5) were seen (57). MLO detection in *Arabidopsis* by electron microscopy is similar to that in other high titer plant species.

DISCUSSION AND CONCLUSIONS

We have shown that the brassicaceous weed *Arabidopsis thaliana* can serve as a host both to the phytopathogens *S. citri* and BLTVA, and to their leafhopper vector *C. tenellus*. Infection of *Arabidopsis* can be confirmed by symptomatology, pathogen cultivation (spiroplasma only), electron microscopy or ELISA. Symptoms of *S. citri* infection appear in as little as 14 days compared to 5 weeks for BLTVA. BLTVA causes only floral symptoms in *Arabidopsis* and it is possible that growth under long days might shorten the latent period. The symptomatology of these two mollicutes in *Arabidopsis* is very different and would provide an interesting comparative study. Other mollicutes such as the aster yellows MLO, which can be transmitted by *Macrosteles* sp. to other brassicas, are likely to infect *Arabidopsis*.

In some ways the small size of *Arabidopsis* plants may be a disadvantage in mollicute experiments, because the effects of leafhopper feeding stress are increased. In addition, the short life span of the plant makes its window of usefulness relatively short. Thus, it is unlikely that *Arabidopsis* would be used as a long-term maintenance host for mollicutes. No resistance to mollicute infection has been found in the small number of *Arabidopsis* ecotypes examined, which precludes investigating host resistance mechanisms and genetics at this time.

One advantage of *Arabidopsis* is size, which permits more plants to be grouped in cages for insect transmission, or the use of smaller cages. Also, the time for symptom appearance is relatively short for *S. citri*, which allows disease evaluation in about two weeks. Although we used 4-7 day inoculation access periods, this period might be shortened; *C. tenellus* transmitted *S. citri* to turnip with inoculation access periods as short as 15 min (37). Although symptom expression requires more time with BLTVA infection, the latent period is short compared to that in other hosts. Another advantage of this system is that *C. tenellus* not only feeds on *Arabidopsis*, but can reproduce on it. In our experience, plant species which are hosts of both pathogen and vector are the most useful for detailed transmission studies.

Each *Arabidopsis* ecotype examined was susceptible to each mollicute to which it was exposed. If resistant ecotypes are found, as with the *Xanthomonas campestris* pv. *campestris* system (153), perhaps the genes for resistance in plant hosts could be identified. Nevertheless, the *Arabidopsis*-mollicute interaction may be a valuable model for investigations of molecular aspects of spiroplasma and MLO infection in plants.

Little is known about the premature induction of flowering in plants infected by BLTVA (56). The discovery that BLTVA can infect all of the biotypes screened thus far suggests that further research on the mechanism of the BLTVA host induction response may be possible using physiological mutants of *Arabidopsis*. Insights into the physiological and genetic reasons for the unique symptoms caused by virescence MLOs in their hosts (Fig. 4) may be gained by utilizing the wide range of developmental mutants having altered flower organs.

Thus, *Arabidopsis* may be an excellent test plant for some mollicute-plant investigations. As our knowledge of the system and the pathogens grows, additional applications and uses of the model will be developed.

BIBLIOGRAPHY

1. Alin, P., Danielson, U. H., and Marverik, P. 1985. 4-Hydroxylalkenals are substrates for glutathione transferase. FEBS Lett. 179:267-270.

2. Apostol, I., Heinstein, P. F., and Low, P. S. 1989. Rapid stimulation of an oxidative burst during the elicitation of cultured plant cells. Plant Physiol. 90:109-116.

3. Arlat, M., Gough, C. L., Barber, C. E., Boucher, C., and Daniels, M. J. 1991. *Xanthomonas campestris* contains a cluster of *hrp* genes related to the larger *hrp* cluster of *Pseudomonas solanacearum*. Mol. Plant-Microbe Interact. 4:593-601.

4. Ausubel, F. M., Brent, R., Kingston R. E., Moore, D. D., Seidman, J. G., Smith, J. A., and Struhl, K. 1991. Current Protocols in Molecular Biology. Greene Publishing Associates/Wiley Interscience, New York.

5. Bailey, J. A., and Mansfield, J. W. 1982. *Phytoalexins*. Blackie, Glasgow, UK.

6. Browne, L. M., Conn, K. L., Ayer, W. A., and Tewari, J. P. 1991. The camalexins: New phytoalexins produced in the leaves of *Camelina sativa* (Cruciferae). Tetrahedron 47:3909-3914.

7. Carrington, J. C., Morris, T. J., Stockley, P. G., and Harrison, S. C. 1987. Structure and assembly of turnip crinkle virus. IV. Analysis of the coat protein gene and implications of the subunit primary structure. J. Mol. Biol. 194:265-276.

8. Carrington, J. C., Keaton, L. A., Zuidema, D., Hillman, B. I., and Morris, T. J. 1989. The complete genome structure of turnip crinkle virus. Virology 170:214-218.

9. Chang, C., Bowman, J. L., DeJohn, A. W., Lander E. S., and Meyerowitz, E. M. 1988. Restriction fragment length polymorphism linkage map for *Arabidopsis thaliana*. Proc. Natl. Acad. Sci. USA 85:6856-6860.

10. Collinge, D. B., and Slusarenko, A. J. 1987. Plant gene expression in response to pathogens. Plant Mol. Biol. 9:389-410.

11. Collmer, C. W., Stenzler, L., Chen, X., Fay, N., Hacker, D., and Howell, S. H. 1992. Single amino acid change in the helicase domain of the putative RNA replicase of turnip crinkle virus alters symptom intensification by virulent satellites. Proc. Natl. Acad. Sci. USA 89:309-313.

12. Cook, W. C. 1967. Life history, host plants, and migration of the beet leafhopper in the western United States. U.S. Dept. Agric. Tech. Bull. 1365.

13. Cretin, C., and Puigdomenech, P. 1990. Glycine-rich RNA-binding proteins from *Sorghum vulgare*. Plant Mol. Biol. 15:783-785.

14. Crute, I. R. 1987. Genetic studies with *Bremia lactucae* (lettuce downy mildew). 207-219 in: Genetics and Plant Pathogenesis. P. R. Day and J. G. Jellis, eds. Blackwell Scientific Publications, Oxford.

15. Crute, I. R., and Norwood, J. M. 1986. Gene dosage effects on the relationship between *Bremia lactucae* (downy mildew) and *Lactuca sativa* (lettuce); the relevance to a mechanistic understanding of host-parasite specificity. Physiol. Plant Path. 29:133-145.

16. Damm, B., Schmidt, R., and Willmitzer, L. 1989. Efficient transformation of *Arabidopsis thaliana* using direct gene transfer to protoplasts. Mol. Gen. Genet. 217:6-12.

17. Dangl, J. L., Holub, E. B., Debener, T., Lehnackers, H., Ritter, C. G., and Crute, I. R. 1992. Genetic definition of loci involved in *Arabidopsis*-pathogen interactions. (in press) in: Methods in *Arabidopsis* Research. C. Koncz, N-H Chua and J. Schell, eds. World Scientific Publishing Co., Singapore.

18. Daniels, M. J., Barber, C. E., Turner, P. C., Cleary S. G., and Sawczyc, K. 1984. Isolation of mutants of *Xanthomonas campestris* pv. *campestris* showing altered pathogenicity. J. Gen. Microbiol. 130:2447-2455.

19. Daniels, M. J., Fan, M. -J., Barber, C. E., Clarke, B. R., and Parker, J. E. 1991. Interaction between *Arabidopsis thaliana* and *Xanthomonas campestris* pp. 84-89 in: Advances in Molecular Genetics of Plant-Microbe Interactions. H. Hennecke and D. P. S. Verma, eds., Vol. 1. Kluwer Academic Publishers, Dordrecht, Netherlands.

20. Daniels, M. J., Markham, P. G., Meddins, B. M., Plaskitt, A. K., Townsend, R., and Bar-Joseph, M. 1973. Axenic culture of

a plant pathogenic Spiroplasma. Nature 244:523-524.

21. Daniels, M. J., Barber, C. E., Turner, P. C., Sawczyc, M. K., Byrde, R. J. W., and Fielding, A. H. 1984. Cloning of genes involved in pathogenicity of *Xanthomonas campestris* pv. *campestris* using the broad host range cosmid pLAFR1. EMBO J. 3:3323-3328.

22. Davis, K. R. 1992. *Arabidopsis thaliana* as a model host for studying plant-pathogen interactions. pp. 393-406 in: Molecular Signals in Plant-Microbe Communication. D. P. S. Verma, ed. CRC Press, Inc., Boca Raton.

23. Davis, K. R., Schott, E., Dong, X., and Ausubel, F. M. 1989. *Arabidopsis thaliana* as a model system for studying plant-pathogen interactions. pp. 99-106 in: Signal Molecules in Plants and Plant-Microbe Interactions. B. J. J. Lugtenberg, ed. Springer-Verlag, Berlin.

24. Davis, K. R., and Ausubel, F. M. 1989. Characterization of elicitor-induced defense responses in suspension-cultured cells of *Arabidopsis*. Mol. Plant Microbe Interact. 2:363-368.

25. Davis, K. R., Schott, E., and Ausubel, F. M. 1991. Virulence of selected phyto-pathogenic pseudomonads in *Arabidopsis thaliana*. Mol. Plant-Microbe Interact. 4:477-488.

26. De Feyter, R., and Gabriel, D. W. 1991. At least six avirulence genes are clustered on a 90-kilobase plasmid in *Xanthomonas campestris* pv. *malvacearum*. Mol. Plant-Microbe Interact. 4:423-432.

27. Debener, T., Lehnackers, H., Arnold, M., and Dangl, J.L. 1991. Identification and molecular mapping of a single *Arabidopsis thaliana* locus determining resistance to a phytopathogenic *Pseudomonas syringae* isolate. Plant J. 1:289-302.

28. Devys, M., Barbier, M., Loiselet, I., Rouxel, T., Sarniguet, A., Kollman, A., and Bousquet, J. -F. 1988. Brassilexin, a novel sulfur-containing phytoalexin from *Brassica juncea* L. (Cruciferea). Tet. Lett. 29:6447-6448.

29. Devys, M., Barbier, M., Kollman, A., Rouxel, T., and Bousquet, J. -F. 1990. Cyclobrassinen sulphoxide, a sulfur-containing phytoalexin from *Brassica juncea*. Phytochemistry 29:1087-1088.

30. Dhawale, S., Souciet, G., and Kuhn, D. N. 1989. Increase of chalcone synthase mRNA in pathogen-inoculated soybeans with race-specific resistance is different in leaves and roots. Plant Physiol. 91:911-916.

31. Dixon, R. A., and Lamb, C. J. 1990. Molecular communication in interactions between plants and microbial pathogens. Annu. Rev. Plant Physiol. Plant Mol. Biol. 41:339-367.

32. Dong, X., Mindrinos, M., Davis, K. R., and Ausubel, F. M. 1991. Induction of *Arabidopsis* defense genes by virulent and avirulent *Pseudomonas syringae* strains and by a cloned avirulence gene. Plant Cell 3:61-72.

33. Dow, J. M., Clarke, B. R., Milligan, D. E., Tang, J. -L., and Daniels, M. J. 1990. Extracellular proteases from *Xanthomonas campestris* pv. *campestris*, the black rot pathogen. Appl. and Env. Microbiol. 56:2994-2998.

34. Dow, J. M., Scofield, G., Trafford, K., Turner, P. C., and Daniels, M. J. 1987. A gene cluster in *Xanthomonas campestris* pv. *campestris* required for pathogenicity controls the excretion of polygalacturonate lyase and other enzymes. Physiol. and Mol. Plant Pathol. 31:261-271.

35. Drews, G. N., Bowman, J. L., and Meyerowitz, E. M. 1991. Negative regulation of the *Arabidopsis* homeotic gene AGAMOUS by the APET- ALA2 product. Cell 65:991-1102.

36. Dreyfuss, G., Philipson, L., and Mattaj, I. W. 1988. Ribonucleoprotein particles in cellular processes. J. Cell. Biol. 106:1419-1425.

37. Eastman, C. E., Schultz, G. A., McGuire, M. R., Post, S. L., and Fletcher, J. 1988. Characteristics of the transmission of a horseradish brittle root isolate of *Spiroplasma citri* by the beet leafhopper *Circulifer tenellus* (Homoptera: Cicadellidae). J. Econ. Entomol. 81:172-177.

38. Ecker, J. R. 1990. PFGE and YAC analysis of the *Arabidopsis* genome. Methods: A Companion to Methods in Enzymology 1(2):186-194.

39. Edwards, K. et al. 1985. Rapid transient induction of phenylalanine ammonia-lyase mRNA in elicitor-treated

bean cells. Proc. Natl. Acad. Sci. USA
82:6731-6735.

40. Ellingboe, A. H. 1982. Genetical aspects of active defense. pp. 179-192 in:
Active Defense Mechanisms in Plants.
R.K.S. Wood, ed. Plenum Press, New York.

41. Errampalli, D., Patton, D., Castle, L.,
Mickelson, L., Hansen, K., Schnall, J.,
Feldmann, K., and Meinke, D. 1991.
Embryonic lethals and T-DNA insertional
mutagenesis in *Arabidopsis*. Plant Cell
3:149-157.

42. Estelle, M. A., and Somerville, C. R.
1986. The mutants of *Arabidopsis*.
Trends in Genetics 2:89-93.

43. Feinbaum, R. L., and Ausubel, F. M.
1988. Transcriptional regulation of the
Arabidopsis thaliana chalcone synthase
gene. Mol. Cell. Biol. 8:1985-1992.

44. Feldmann, K. A. 1991. T-DNA insertion
mutagenesis in *Arabidopsis* - mutational
spectrum. Plant J. 1:71-82.

45. Feldmann, K. A., Carlson, T. J., Coomber,
S. A., Farrance, C. E., Mandel, M. A.,
and Wierzbicki, A. M. 1990. T-DNA
insertional mutagenesis in *Arabidopsis
thaliana*. Hort. Biotech. 11:109-120.

46. Fellay, R., Rahme, L. G., Mindrinos, M.
N., Frederick, R. D., Pisi, A., and
Panopoulos, N. P. 1991. Genes and signals controlling the *Pseudomonas syringae*
pv. *phaseolicola*-plant interaction. pp.
45-52 in: Advances in Molecular Genetics
of Plant-Microbe Interactions. Vol. 1,
H. Hennecke and D. P. S. Verma, eds.
Kluwer Academic Publishers, Dordrecht,
Netherlands.

47. Figurski, D., and Helinksi, D. R. 1979. Replication of an origin-containing derivative of plasmid RK2 dependent on a plasmid function provided in trans. Proc. Natl. Acad. Sci. USA. 76:1648-1652.

48. Fletcher, J., Schultz, G. A., Davis, R. E., Eastman, C. E., and Goodman R. M. 1981. Brittleroot disease of horseradish: Evidence for an etiological role of *Spiroplasma citri*. Phytopathology 71:1073-1080.

49. Fletcher, J., and Eastman, C. E. 1984. Translocation and multiplication of *Spiroplasma citri* in turnip (*Brassica rapa*). Curr. Microbiol. 11:289-292.

50. Fletcher, J., and Eastman, C. E. 1991. *Arabidopsis thaliana* as an experimental host plant of *Spiroplasma citri*. Phytopathology 81:1209.

51. Flor, H. 1971. Current status of the gene-for-gene concept. Annu. Rev. Phytopathol. 9:275-296.

52. Fraser, R. S. S. 1990. The genetics of resistance to plant viruses. Annu. Rev. Phytopathol. 28:179-200.

53. Gabriel, D. W., Burges, A., and Lazo, G. R. 1986. Gene-for-gene recognition of five cloned avirulence genes from *Xanthomonas campestris* pv. *malvacearum* by specific resistance genes in cotton. Proc. Natl. Acad. Sci. USA 83:6415.

54. Gabriel, D. W., and Rolfe, B. G. 1990. Working models of specific recognition in plant-microbe interactions. Annu. Rev. Phytopathol. 28:365-391.

55. Golino, D. A., Oldfield, G. N., and Gumpf, D. J. 1987. Transmission characteristics of the beet leafhopper transmitted virescence agent (BLTVA). Phytopathology 77:954-957.

56. Golino, D. A., Oldfield, G. N., and Gumpf, D. J. 1988. Induction of flowering through infection by beet leafhopper transmitted virescence agent. Phytopathology 77:954-957.

57. Golino, D. A., Shaw, M., and Rappaport, L. 1988. Infection of *Arabidopsis thaliana* (L.) Heynh. with a mycoplasmalike organism, the beet leafhopper transmitted virescence agent. Arabidopsis Information Serv. 26:9-14.

58. Golino, D. A., Oldfield, G. N., and Gumpf, D. G. 1989. Experimental hosts of the beet leafhopper transmitted virescence agent. Plant Disease 73:850-854.

59. Golino, D. A., Kirkpatrick, B. C., and Fisher, G. 1989. The production of a polyclonal antisera to the beet leafhopper transmitted virescence agent. Phytopathology 79:1138.

60. Gomez, J., Sanchez-Martinez, D., Stiefel, V., Rigau, J., Puigdomenech, P., and Pages, M. 1988. A gene induced by the plant hormone abscisic acid in response to water stress encodes a glycine-rich protein. Nature 334:262-264.

61. Gough, C. L., Dow, J. M., Barber, C. E., and Daniels, M. J. 1988. Cloning of two endoglucanase genes of *Xanthomonas campestris* pv. *campestris*: analysis of the role of the major endoglucanase in

pathogenesis. Mol. Plant-Microbe Interact. 1:275-281.

62. Graham, T. L., and Graham, M. Y. 1991.
Cellular coordination of molecular re-
sponses in plant defense. Mol.
Plant-Microbe Interact. 4:415-422.

63. Griffing, B., and Scholl, R. L. 1991.
Qualitative and quantitative genetic
studies of *Arabidopsis thaliana*. Genetics 129:605-609.

64. Grill, E., and Somerville, C. 1991.
Construction and characterization of a
yeast artificial chromosome library of
Arabidopsis which is suitable for chromosome walking. Mol. Gen. Genet.
226:484-490.

65. Habereder, H., Schroder, G., and Ebel, J.
1989. Rapid induction of phenylalanine
ammonia-lyase and chalcone synthase mRNAs
during fungus infection of soybean *(Glycine max* L.) roots or elicitor treatment
of soybean cell cultures at the onset of
phytoalexin synthesis. Planta 177:58-65.

66. Hahlbrock, K., and Scheel, D. 1989.
Physiology and molecular biology of
phenylpropanoid metabolism. Annu. Rev.
Plant Physiol. Plant Mol. Biol.
40:347-369.

67. Hauge, B. M., Hanley, S., Giraudat, J.,
and Goodman, H. M. 1992. Mapping the
Arabidopsis genome. (in press). in:
Molecular Biology of Plant Development.
G. Jenkins and W. Schurch, eds. Cambridge
University Press, Cambridge.

68. Heaton, L. A., Carrington, J. C., and
Morris, T. J. 1989. Turnip crinkle

virus infection from RNA synthesized *in vitro*. Virology 170:214-218.

69. Hennebert, G. L., and Korf, R. P. 1975. The peat mould, *Chromelosporium ollare*, conidial state of *Peziza ostracoderma*, and its misapplied names, *Botrytis crystallina*, *Botrytis spectabilis*, *Ostracoderma epigaeum* and *Peziza atrovinosa*. Mycologia 67:214-240.

70. Hitchin, F. E., Jenner, C., Harper, S., Mansfield, J., Barber, C., and Daniels, M. 1989. Determinant of cultivar specific avirulence cloned from *Pseudomonas syringae* pv. *phaseolicola* race 3. Physiol. Mol. Plant Pathol. 34:309-322.

71. Hoisington, D. A., and Coe, E. H. J. 1989. Methods for correlating RFLP maps with conventional genetic and physical maps in maize. pp. 19-24 in: Current Communications in Plant Molecular Biology: Development and Application of Molecular Markers to Problems in Plant Genetics. T. Helentjarius and B. Burr, eds. Cold Spring Harbor Laboratory Press, New York.

72. Huynh, T., Dahlbeck, D., and Staskawicz, B.J. 1989. Bacterial blight of soybeans: Regulation of a pathogen gene determining host cultivar specificity. Science 245:1374-1377.

73. Hwang, I., Kohchi, T., Hauge, B. M., Goodman, H. M., Schmidt, R., Cnops, G., Dean, C., Gibson, S., Iba, K., Lemieux, B., Arondel, V., Danhoff L., and Somerville, C. 1991. Identification and map position of YAC clones comprising one-third of the *Arabidopsis* genome. Plant J. 1:367-374.

74. Jefferson, R. A., Kavanagh, T. A.and Bevan, M. W. 1987. GUS fusions: Betaglucuronidase as a sensitive and versatile gene fusion marker in higher plants. EMBO J. 6:3901.

75. Jejelowo, O. A., Conn, K. L., and Tewari, J. P. 1991. Relationship between conidial concentration, germling growth, and phytoalexin production by Camelina sativa leaves inoculated with Alternaria brassicae. Mycol. Res. 95:928-934.

76. Jenner, C., Hitchin, E., Mansfield, J., Walters, K., Betteridge, P., Teverson, D. and Taylor, J. 1991. Gene-for-gene interactions between Pseudomonas syringae pv. phaseolicola and Phaseolus. Mol. Plant-Microbe Interact. 4:553-562.

77. Keen, N. T., and Staskawicz, B. 1988. Host range determinants in plant pathogens and symbionts. Annu. Rev. Microbiol. 42: 421.

78. Keen, N. T. 1990. Gene-for-gene complementarity in plant-pathogen interactions. Annu. Rev. Genet. 24:447-463.

79. Keen, N. T., Tamaki, S., Kobayashi, D., Gerhold, D., Stayton, M., Shen, H., Gold, S., Lorang, J., Thordal-Christensen, H., Dahlbeck, D., and Staskawicz, B. 1990. Bacteria expressing avirulence gene D produce a specific elicitor of the soybean hypersensitive reaction. Mol. Plant-Microbe Interact. 3:112-121.

80. Keen, N. T., and Buzzel, R. I. 1991. New disease resistance genes in soybean against Pseudomonas syringae pv. glycinea: evidence that one of them interacts with a bacterial elicitor. Theor. Appl. Genet. 81:133-138.

81. Keith, B., Dong, X., Ausubel, F. M., and Fink, G. R. 1991. Differential induction of 3-deoxy-D-heptulosonate 7-phosphate synthase gene in *Arabidopsis thaliana* by wounding and pathogen attack. Proc. Natl. Acad. Sci. USA 88:8821-8825.

82. King, E. O., Ward, M. K., and Raney, D. E. 1954. Two simple media for the demonstration of phycocyanin and fluorescin. J. Lab. Clin. Med. 44:301-307.

83. Kirkpatrick, B. C. 1989. Strategies for characterizing plant pathogenic mycoplasma-like organisms and their effects on plants. pp. 241-293 in: Microbe Interactions, Molecular and Genetic Perspectives; vol 3. Kosuge and Nester, eds. McGraw-Hill, Inc. New York.

84. Kobayashi, D. Y., Tamaki, S. J., and Keen, N. T. 1989. Cloned avirulence genes from the tomato pathogen *Pseudomonas syringae* pv. *tomato* confer cultivar specifity on soybean. Proc. Natl. Acad. Sci. USA 86:157-161.

85. Kobayashi, D. Y., Tamaki, S. J., and Keen, N. T. 1990. Molecular characterization of avirulence gene D from *Pseudomonas syringae* pv. *tomato*. Mol. Plant-Microbe Interact. 3:94-102.

86. Koch, E., and Slusarenko, A. J. 1990a. Fungal pathogens of *Arabidopsis* (L.) Heynh. Botanica Helvetica 100:257-268.

87. Koch, E., and Slusarenko, A. J. 1990b. *Arabidopsis* is susceptible to infection by a downy mildew fungus. Plant Cell 2:437-445.

88. Koornnef, M. 1987. Linkage map of *Arabidopsis thaliana* (2n =10). pp.

742-745 in: Genetic Maps. S. J. O'Brien, ed. Cold Spring Harbor Laboratory Press, Cold Spring Harbor.

89. Kranz, A. R., and Kirchheim, B. 1987. Genetic resources in *Arabidopsis*. Arabidopsis Inf. Serv. 24:1-386.

90. Kuc, J. 1972. Phytoalexins. Annu. Rev. Phytopathol. 10:207-232.

91. Kustu, S., Santero, E., Keener, J., Popham D., and Weiss, D. 1989. Expression of σ^{54} *(ntrA)*-dependent genes is probably united by a common mechanism. Microbiol. Rev. 53:367-376.

92. Laibach, F. 1943. *Arabidopsis thaliana* (L.) Heynh. als Objekt für genetische und entwicklungs-physiologische Untersuchungen. Botanisches Archiv 44:439-455.

93. Lamb, C. J., Lawton, M. A., Dron, M., and Dixon, R. A. 1989. Signals and transduction mechanisims for activation of plant defenses against microbial attack. Cell 56:215-224.

94. Lander, E. R., and Botstein, D. 1986. Strategies for studying heterologous traits in humans by using a linkage map of restriction fragment length polymorphisms. Proc. Natl. Acad. Sci. USA 83:7353-7357.

95. Langridge, J. 1955. Biochemical mutations in the crucifer *Arabidopsis thaliana* (L.) Heynh. Nature 176:260-261.

96. Leutwiler, L. S., Hough-Evans, B. R., and Meyerowitz, E. M. 1984. The DNA of *Arabidopsis thaliana*. Mol. Gen. Genet. 194:15-23.

97. Li, X. H., Heaton, L., Morris, T. J., and Simon, A. E. 1989. Defective inter-fering RNAs of turnip crinkle virus in-tensify viral symptoms and are generated *de novo*. Proc. Natl. Acad. Sci. USA 86:9173-9177.

98. Li, X. H., and Simon, A. E. 1990. Symptom intensification on cruciferous hosts by the virulent sat-RNA of turnip crinkle virus. Phytopathology 80:238-242.

99. Li, X. H., and Simon, A. E. 1991. *In vivo* accumulation of a turnip crinkle virus DI RNA is affected by alterations in size and sequence. J. Virol. 65:4582-4590.

100. Liang, X., Dron, M., Cramer, C. L., Dixon R. A., and Lamb, C. J. 1989. Differential regulation of phenylalanine ammonia-lyase genes during plant devel-opment and by environmental clues. J. Biol. Chem. 254:14486-14492.

101. Liu, Y. -N., Tang, J. -L., Clarke, B. R., Dow, J. M., and Daniels, M. J. 1990. A multipurpose broad host range cloning vector and its use to characterize an extracellular protease gene of *Xanthomonas campestris* pv. *campestris*. Mol. Gen. Genet. 220:433-440.

102. Lois, R., Dietrich, A., Hahlbrock, K., and Schulz, W. 1989. A phenylalanine ammonia-lyase gene from parsley: struc-ture, regulation and identification of elicitor and light responsive cis-acting elements. EMBO J. 8:1641-1648.

103. Maniatis, T., Fritsch, E. F., and Sambrook, J. 1982. Molecular Cloning: A Laboratory Manual. Cold Spring Harbor Laboratory, Cold Spring Harbor, New York.

104. Markham, P. G., and Townsend, R. 1979. Experimental vectors of spiroplasmas. pp. 413-446 in: Leafhopper Vectors and Plant Disease Agents. K. Maramorosch and K. F. Harris, eds. Academic Press, New York.

105. Marks, M. D., and Feldmann, K. A. 1989. Trichome development in *Arabidopsis thaliana*. 1. T-DNA tagging of the glabrous1 gene. Plant Cell 1:1043-1050.

106. Márton, L., and Browse, J. 1991. Facile transformation of *Arabidopsis*. Plant Cell Rep. 10:235-239.

107. Matthews, R. E. F. 1991. Plant Virology. Third edition, Academic Press, Inc., San Diego, California.

108. McCourt, P., and Somerville, C. R. 1987. The use of mutants for the study of plant metabolism. pp. 33-64 in: The Biochemistry of Plants. D. D. Davies, ed. Academic Press, San Diego.

109. McCoy, R. E., Caudwell, A., Chang, C. J., Chen, T. A., Chiykowski, L. N., Cousin, M. T., Dale, J. L., DeLeeuw, G. T. N., Golino, D. A., Hackett, K. J., Kirkpatrick, B. C., Marwitz, R., Petzold, H., Sinha, R. C., Sugiura, M., Whitcomb, R. F., Yang, I. L., Zhu, B. M., and Seemuller, E. 1989. Plant diseases associated with mycoplasmalike organisms. pp. 545-640 in: The Mycoplasmas; Vol 5. R. F. Whitcomb and J.

G. Tully, eds. Academic Press, Inc., New York.

110. Melcher, U., Gardner, C. E. Jr., and Essenberg, R. C. 1981. Clones of cauliflower mosaic virus identified by molecular hybridization in turnip leaves. Plant Mol. Biol. 1:63-73.

111. Métraux, J. P., Ahl Goy, P., Staub, Th., Speich, J., Steinmann, A., Ryals, J., and Ward, E. 1991. Induced systemic resistance in cucumber in response to 2,6-dichloro-isonicotinic acid and pathogens, pp. 432-439 in: Advances in Molecular Genetics of Plant-Microbe Interactions. H. Hennecke and D. P. S. Verma, eds. Kluwer Academic Publishers, Dordrecht, The Netherlands.

112. Meyerowitz, E. M. 1987. *Arabidopsis thaliana*. Annu. Rev. Genet. 21:93-111.

113. Meyerowitz, E. M. 1989. *Arabidopsis*, a useful weed. Cell 56:263-269.

114. Monde, K., Sasaki, K., Shirata, A., and Takasugi, M. 1990. 4-Methoxybrassinen, a sulfur containing phytoalexin from *Brassica oleracea*. Phytochemistry 29:1499-1500.

115. Monde, K., Sasaki, K., Shirata, A., and Takasugi, M. 1991. Methoxybrassenins A and B, sulphur containing stress metabolites from *Brassica oleracea* var. capitata. Phytochemistry 30:3921-3922.

116. Monde, K., Sasaki, K., Shirata, A., and Takasugi, M. 1991. Brassicanal C and two dioxindoles from cabbage. Phytochemistry 30:2915-2917.

117. Muller, K. O., and Borger, H. 1940. Experimentelle unersuchengen uber die *Phytophthora*-resistenz den kartoffel. Arb. Biol. Bund. Land. Forst. (Berl). 23:189-231.

118. Muller, K. 1956. Einege einfache versuchenzum nachweiss von phytoalexinen. Phytopath. Z. 27:237-254.

119. Nam, H. -G., Giraudat, J., den Boer, B., Moonan, F., Loos, W. D. B., Hauge, B. M., and Goodman, B. M. 1989. Restriction fragment length polymorphism linkage map of *Arabidopsis thaliana*. Plant Cell 1:699-705.

120. Oldfield, G. N., and Kaloostian, G. H. 1979. Vectors and host range of the citrus stubborn disease pathogen, *Spiroplasma citri*. pp. 119-124 in: Proc. R.O.C.-U.S. Coop. Sci. Seminar on Mycoplasma Diseases of Plants. (NSC Symp. Ser. 1) Nat. Sci. Counc. Rep. of China, Taiwan.

121. Osbourn, A. E., Clarke, B. R., and Daniels, M. J. 1990. Identification and DNA sequence of a pathogenicity gene of *Xanthomonas campestris* pv. *campestris*. Mol. Plant- Microbe Interact. 3:280-285.

122. Paxton, J. D. 1981. Phytoalexins-A working redefinition. Phytopath. Z. 101:106-109.

123. Rahme, L. G. 1991. The hrp ("Harp") Genes of *Pseudomonas syringae* pv. *phaseolicola*: Organization, Transcription, Signalling and Role in the Plant-Bacterium Interactions, Ph.D.

Thesis, University of California, Berkeley.

124. Redei, G. P. 1975. *Arabidopsis* as a genetic tool. Annu. Rev. Genet. 9:111.

125. Reinholz, E. 1947. Auslösung von röntgenmutationen bei *Arabidopsis thaliana* (L.) Heynh. und ihre bedeutung für die pflanzenzuchtung and evolutionstheorie. Field Inform. Agency Tech. Rep. 1006:1-70.

126. Reiter, R. S., Williams, J., G. K., Feldmann, K. A., Rafalski, J. A., Tingey, S. V., and Scolnik, P. A. 1992. Global and local genome mapping in *Arabidopsis thaliana* by using recombinant inbred lines and random amplified polymorphic DNAs. Proc. Natl. Acad. Sci. USA 89: 1477- 1481.

127. Reitzer, L. J., and Magasanick, B. 1989. Activation of glnA in *E. coli* is stimulated by activator bound to sites far from the promoter. Cell 45:785-792.

128. Rouxel, T., Sarniguet, A., Kollman, A., and Bousquet, J. -F. 1989. Accumulation of a phytoalexin in *Brassica* spp. in relation to a hypersensitive reaction to *Leptosphaeria maculans*. Physiol. Mol. Plant Pathol. 34:507-514.

129. Rouxel, T., Renard, M., Kollman, A., and Bousquet, J. -F. 1990. Brassilexin accumulation and resistance to *Leptoshaeria maculans* in *Brassica* spp., and progeny of an interspecific cross *B. junceae* X *B. napus*. Euphytica 46:175-181.

130. Rouxel, T., Kollman, A., Boulidard, L., and Mithen, R. 1991. Abiotic elicita-

tion of indole phytoalexins and resistance to *Leptosphaeria maculans* within Brassicaceae. Planta 184:271-278.

131. Ruvkun, G. B., and Ausubel, F. M. 1981. A general method for site directed mutagenesis in prokaryotes. Nature 289:85-88.

132. Samac, D. A., Hironaka, C. M., Yallaly, P. E., and Shah, D. M. 1990. Isolation and characterization of the genes encoding basic and acidic chitinase in *Arabidopsis thaliana*. Plant Physiol. 93:907-914.

133. Samac, D. A., and Shah, D. M. 1991. Developmental and pathogen-induced activation of the *Arabidopsis* acidic chitinase promoter. Plant Cell 3:1063-1072.

134. Sambrook, J., Fritsch, E. F., and Maniatis, T. 1989. Molecular Cloning: A Laboratory Manual. Cold Spring Harbor Laboratory, Cold Spring Harbor, N.Y.

135. Sangwan, R. S., Bourgeois, Y., Sangwannorreel, B. S. 1991. Genetic transformation of *Arabidopsis thaliana* zygotic embryos and identification of critical parameters influencing transformation efficiency. Mol. Gen. Genet. 230:475-485.

136. Seki, M., Shigemoto, N., Komeda, Y., Imamura, J., Yamada, Y., and Morikawa, H. 1991. Transgenic *Arabidopsis thaliana* plants obtained by particle-bombardment-mediated transformation. Appl. Microbiol. Biotechnol. 36:228-230.

137. Simmonds, N. W. 1991. Genetics of horizontal resistance to diseases of crops. Biol. Rev. 66:189-241.

138. Simon, A. E., and Howell, S. H. 1986. The virulent satellite RNA of turnip crinkle virus has a major domain homologous to the 3'-end of the helper virus genome. EMBO J. 5:3423-3428.

139. Simpson, R. B., and Johnson, L. J. 1990. *Arabidopsis thaliana* as a host for *Xanthomonas campestris* pv. *campestris*. Mol. Plant-Microbe Interact. 3:233-237.

140. Slusarenko, A. J., and Mauch-Mani, B. 1991. Downy mildew of *Arabidopsis thaliana* caused by *Peronospora parasitica*: a model system for the investigation of the molecular biology of host-pathogen interactions. pp. 280-283 in: Advances in Molecular Genetics of Plant-Microbe Interactions Vol. 1. H. Hennecke and D. P. S. Verma, eds. Kluwer Academic Publishers, Dordrecht.

141. Somerville, C. 1989. *Arabidopsis* blooms. Plant Cell 1:1131-1135.

142. Staskawicz, B. J., Dahlbeck, D., and Keen, N. T. 1984. Cloned avirulence gene of *Pseudomonas syringae* pv. *glycinea* determines race-specific incompatibility of *Glycine max* (L.) Merr. Proc. Natl. Acad. Sci. USA 81:6024-6028.

143. Staskawicz, B., Dahlbeck, D., Keen N., and Napoli, C. 1987. Molecular characterization of cloned avirulence genes from race 0 and race 1 of *Pseudomonas syringae* pv. *glycinea*. J. Bacteriol. 169:5789-5794.

144. Straus, D., and Ausubel, F. M. 1990. Genomic subtraction for cloning DNA corresponding to deletion mutations. Proc. Natl. Acad. Sci. USA 87:1889-1893.

145. Sun, T. -P., Goodman, H. M., and Ausubel, F. M. 1992. Cloning the *Arabidopsis GA-1* locus by genomic subtraction. Plant Cell 4:119-128.

146. Swanson, J., Kearney, B., Dahlbeck, D., and Staskawicz, B. J. 1988. Cloned avirulence gene of *Xanthomonas campestris* pv. *vesicatoria* complements spontaneous race change mutant. Mol. Plant-Microbe Interact. 1:5-9.

147. Takasugi, M., Monde, K., Katsui, N., and Shirati, A. 1987. Spirobrassinen, a novel sulphur- containing phytoalexin from the daikon *Raphanus sativus* 1. var. *hortensis* (Cruciferae) Chem. Lett. 1631-1632.

148. Takasugi, M., Monde, K., Katsui, N., and Shirata, A. 1988. Novel sulfur-containing phytoalexins from the chinese cabbage *Brassica campestris* L. spp. *pekinensis* (Cruciferae). Bull. Chem. Soc. Japan. 61:285-289.

149. Tamaki, S., Dahlbeck, D., Staskawicz, B., and Keen, N. T. 1988. Characterization and expression of two avirulence genes cloned from *Pseudomonas syringae* pv. *glycinea*. J. Bacteriol. 170:4846-4854.

150. Tang, J. -L., Gough, C. L., and Daniels, M. J. 1990. Cloning of genes involved in negative regulation of production of extracellular enzymes and polysaccharide of *Xanthomonas campestris* pv.

campestris. Mol. Gen. Genet. 222:157-160.

151. Tang, J. -L., Liu, Y. -N., Barber, C. E., Dow, J. M., Wootton, J. C., and Daniels, M. J. 1991. Genetic and molecular analysis of a cluster of *rpf* genes involved in positive regulation of synthesis of extracellular enzymes and polysaccharide in *Xanthomonas campestris* pv. *campestris.* Mol. Gen. Genet. 226:409-417.

152. Tegtmeier, K., and Van Etten, H.D. 1982. The role of pistin tolerance and degradation in the virulence of *Nectria haematococca* on pea. Phytopathology 72:608-612.

153. Tsuji, J., Somerville, S. C., and Hammerschmidt, R. 1991. Identification of a gene in *Arabidopsis thaliana* that controls resistance to *Xanthomonas campestris* pr. *campestris.* Physiol. Mol. Plant Path. 38:57-65.

154. Tsuji, J., Jackson, E. P., Gage, D. A. Hammerschmidt, R., and Somerville, S. C. 1992. Phytoalexin accumulation in *Arabidopsis thaliana* during the hypersensitive response to *Pseudomonas syringae* pv. *syringae.* Plant Physiol. 98:1304-1309.

155. Turner, P., Barber, C., and Daniels, M. 1984. Behaviour of the transposons Tn5 and Tn7 in *Xanthomonas campestris* pv. *campestris.* Mol. Gen. Genet. 195:101-107.

156. Turner, P., Barber, C., and Daniels, M. 1985. Evidence for clustered pathogenicity genes in *Xanthomonas campestris*

pv. *campestris.* Mol. Gen. Genet. 199:338-343.

157. Uknes, S., Mauch-Mani, B., Moyer, M., Williams, S., Dincher, S., Chandler, D., Slusarenko, A. J., Ward, E., and Ryals, J. 1992. Gene expression associated with induced resistance in *Arabidopsis thaliana.* Plant Cell 4:645-656.

158. Valvekens, D., Vanmontagu, M., and M. Vanlijsebettens. 1988. *Agrobacterium tumefaciens*-mediated transformation of *Arabidopsis thaliana* root explants by using kanamycin selection. Proc. Natl. Acad. Sci. USA 85:5536-5540.

159. Vanderplank, J. E. 1978. Genetic and molecular basis of plant pathogenesis. Advanced Series in Agricultural Sciences, No. 6, Springer-Verlag.

160. Vankan, J. A. L., Vandenackerveken, G. F. J. M., and Dewit, P. J. G. M. 1991. Cloning and characterization of cDNA of avirulence gene avr9 of the fungal pathogen *Cladosporium fulvum,* causal agent of tomato leaf mold. Mol. Plant-Microbe Interact. 4:52-59.

161. Vivian, A., Atherton, G., Bevan, J., Crute, I., Mur, L., and Taylor, J. 1989. Isolation and characterization of cloned DNA conferring specific avirulence in *Pseudomonas syringae* pv. *pisi* to pea (*Pisum sativum*) cultivars, which posses the resistance allele, R2. Physiol. Mol. Plant. Pathol. 34:335-344.

162. Ward, E. R., and Jen, G. C. 1990. Isolation of single-copy-sequence clones from a yeast artificial chromosome library of randomly-sheared *Arabidopsis*

thaliana DNA. Plant Mol. Biol. 14:561-568.

163. Wei, N., Heaton, L. A., Morris, T. J., and Harrison, C. S. 1990. Structure and assembly of turnip crinkle virus. VI. Identification of coat protein binding sites on the RNA. J. Mol. Biol. 214:85-95.

164. Whalen, M. C., Innes, R. W., Bent, A. F., and Staskawicz, B. J. 1991. Identification of *Pseudomonas syringae* pathogens of *Arabidopsis* and a bacterial locus determining avirulence on both *Arabidopsis* and soybean. Plant Cell 3:49-59.

165. Williams, P. H. 1980. Black rot: a continuing threat to world crucifers. Plant Disease 64:736-742.

166. Willis, D. K., Rich, J. J., and Hrabak, E. M. 1991. *Hrp* genes of phytopathogenic bacteria. Mol. Plant-Microbe Interact. 4:132-138.

167. Woods, A. M., Fagg, J., and Mansfield, J. W. 1988. Fungal development and irreversible membrane damage in cells of *Lactuca sativa* undergoing the hypersensitive reaction to the downy mildew fungus *Bremia lactucae*. Physiol. Mol. Plant Path. 32:483-497.

168. Yanofsky, M. F., Ma, H., Bowman, J. L., Drews, G. N., Feldmann, K. A., and Meyerowitz, E. M. 1990. The protein encoded by the *Arabidopsis* homeotic gene *agamous* resembles transcription factors. Nature 346:35-39.

169. Young, N. D. 1990. Potential applications of map-based cloning to plant

pathology.   Physiol. Mol. Plant Pathol.
37:81-94.

170.   Zhang, C., Cascone, P. J., and Simon, A.
E.   1991.   Recombination between satellite and genomic RNAs of turnip crinkle
virus. Virology 184:791-794.